D0955951

Zero Inventories

658.787
H146

Zero Inventories

Robert W. Hall
Indiana University
with
American Production & Inventory Control Society

WITHDRAWN
ST. MARYS COLLEGE LIBRARY

171502

1983
DOW JONES-IRWIN
Homewood, Illinois 60430

LIBRARY ST. MARY'S COLLEGE

Szf + 5/88

© DOW JONES-IRWIN, 1983

All rights reserved. No part of this publication may be reproduced, stored in a retrieval system, or transmitted, in any form or by any means, electronic, mechanical, photocopying, recording, or otherwise, without the prior written permission of the publisher.

This publication is designed to provide accurate and authoritative information in regard to the subject matter covered. It is sold with the understanding that the publisher is not engaged in rendering legal, accounting, or other professional service. If legal advice or other expert assistance is required, the services of a competent professional person should be sought.
From a Declaration of Principles jointly adopted by a Committee of the American Bar Association and a Committee of Publishers.

ISBN 0-87094-461-4

Library of Congress Catalog Card No. 83–71736

Printed in the United States of America

7 8 9 0 MP 0 9 8 7 6 5

Foreword

The North American economy is facing one of its most serious challenges in overcoming a persistent and innovative foreign competition for consumer goods and services. Of all the foreign competitors, Japan has been the most successful. The Japanese have taken a large share of the North American market from North American manufacturers, and they have been able to beat North American companies on their own home field. This is a significant accomplishment when one considers the many obstacles that a foreign competitor must overcome to achieve this result. The higher cost of transporting the product to the marketplace is one; establishing a distribution network is another. The needs and desires of the customer in another culture and environment must be accurately evaluated in order to design a competitive product. The Japanese succeeded. They succeeded with products manufactured in high-volume, repetitive operations, a segment of industry in which North America was the recognized leader.

Japan is an overpopulated nation scattered over several islands in the Western Pacific. It has very few natural resources and must import most of its raw materials and energy-producing goods. It suffered severe damage in World War II, which ended with the most devastating blow of all—defeat.

The postwar leaders of Japan recognized their problems. They needed hard currency to pay for the imports they had to procure if they were to have a viable economy. The only way to acquire this currency was to export products to foreign markets and become competitive in those markets. The market segment they targeted was that of high-volume consumer products, as this offered the best opportunity to generate the amount of foreign exchange needed. They addressed the issue of producing these products at a cost that would allow them to compete successfully in foreign markets. Government, business, and labor all recognized the challenge and mounted a concerted effort to meet it.

The results are history. North American business leaders are now focusing their attention on Japanese methods. Conducting tours of Japanese manufacturers has become a growth industry in North America. Articles have been written and seminars conducted on Japanese techniques. Much of this attention is focused on production and inventory control.

The leading professional society in this field is the American Production and Inventory Control Society (APICS). For the 25 years of its existence, this organization has identified as its objective the education of practitioners in the body of knowledge of the field of production and inventory control. To satisfy this role, APICS has consistently applied itself to the task of defining (and redefining) the contents of this body of knowledge. It has recognized its responsibility to acquire an understanding of the principles and methods being employed by the Japanese. To accomplish this, a careful, analytic approach was required, one that examines broadly and deeply all facets of Japanese manufacturing.

This examination revealed that, while Japanese manufacturing had applied some innovative approaches to production and inventory control, there were many other factors involved. Product design, process planning, quality assurance, preventive maintenance, and management all contributed to the Japanese success story. They had accomplished much more than excellence in these specific disciplines; they had succeeded in integrating them so they complemented one another in the achievement of company objectives.

Professor Robert W. Hall is deeply involved in the APICS effort. He is a Professor at Indiana University, School of Business. He has been active in APICS, especially in the repetitive manufacturing area. He was a member of the original steering committee that organized production and inventory control practitioners who worked in the repetitive manufacturing industry. He has made several trips to Japan and has maintained a constant dialogue with members of the Japanese industrial and academic communities. He is the author of *Driving the Productivity Machine: Production and Inventory Control in Japan* and *Kawasaki USA—A Case Study,* both published by the American Production and Inventory Control Society.

Bob Hall has not only studied Japanese manufacturers but has worked closely with American companies in an effort to transfer the technology developed by the Japanese to American business. This book contains the results of his extensive observations and analysis of the subject. It covers the many functions that must be orchestrated in order to realize the benefits that can be achieved. He is careful to look beyond the techniques the Japanese applied to the principles upon which they were based. As a result, the application of these techniques to North American business is more practical. The book will benefit all segments of manufacturing, not just high-volume, repetitive companies, for the philosophies driving successful Japanese companies can be applied to all business organizations.

The research and preparation for writing this book was supported by the APICS Educational and Research Foundation, Inc., a separate corporation founded by APICS and dedicated to sponsoring worthwhile projects, such as research and the development of educational materials, that will contribute to the body of knowledge in the field of production and inventory control.

The efforts of APICS to learn about Japanese manufacturing methods and to convert them to applications in North American business have been directed by the board of directors of that organization. This governing body of APICS consists of 24 members elected by its more than 51,000 members in North America. The board has been both sensitive and responsive to the necessity of aggressively applying resources to the analysis of Japanese technology. The dramatic results the Japanese have achieved in reducing inventory levels has stimulated APICS to spearhead the effort to move towards a "Zero Inventories" concept. Currently, the Society is organizing a campaign to make the lessons learned from the Japanese available to North American business. This book is part of that campaign.

I highly recommend this book. It can provide you with new ideas and concepts that can be of benefit to you and your company in competing in what has become an international marketplace.

J. P. Kelleher, CPIM*
President—1982
American Production and
Inventory Control Society

Preface

The goal of this book is to start manufacturing managers toward understanding and applying stockless production. The book certainly is not comprehensive because the subject is new both to the author and to American manufacturers generally. If those who have a serious interest in applying stockless production find it to be a "starter kit," it will serve its purpose. The experienced practitioner of stockless production will find it elementary.

Presenting stockless production in a logical order is very difficult. Almost every aspect of stockless production interacts with every other aspect, so that discussion of it in any depth tends to bounce from topic to topic. Some of the chapters may stand alone, but a complete story is told only by reading the book from front to back. Those with a special interest in stockless production concepts applied to relationships with suppliers will find the subject touched upon in many different places, for example.

Stockless production is simple and not hard to understand. However, it permeates every activity in manufacturing, so that understanding all of it requires consideration of many different specialized topics. A new vocabulary is necessary to express some of the ideas. Despite attempts by the American Production and Inventory Control

Society (and others), there has been no standardization of terms from plant to plant within the United States, and the same is true in Japan. It is particularly difficult to apply suitable terminology to ideas that are not in widespread practice in the United States. A glossary at the end of Chapter 1 will be of some help, but it is impossible to use the terms consistently.

Source material for this effort was gathered over a period of time but is still not adequate. The Japanese auto industry has integrated ideas from many sources into their version of stockless production. Since they pioneered in this, much of the material is from them. Not as much data were available from other industries. I especially regret that more could not be accumulated on applications in the electronics industry.

So many people have contributed to this work that it is impossible to name all of them. It is the result of visits and discussions with many manufacturing executives both in the United States and Japan who gave generously of their time to explain their practices and their industry: Edward Hay, John Kinsey, Larry Higgason, Kichiro Ando, Leonard Ricard, Susumu Okada, Yoshinori Yamada, Ken Wantuck, Leighton Smith, Roy Harmon, David Nelleman, Masukatsu Mori, Fujio Cho, Ed Reznicek, William Sandras. . . . I apologize to all those for whom time and space did not allow a listing.

Jinichiro Nakane deserves special thanks. He first introduced me to the ideas about five years ago and has patiently corrected my misunderstandings ever since. He has arranged visits, discussed the ideas, and illustrated their use, and through him we have had useful exchanges with the Japan Production and Inventory Control Society.

The book is much better for the efforts of four persons who reviewed this book and corrected oversights and contributed suggestions which were used:

John Kinsey, Hoover Universal.

Edmund Reznicek, Delco Manufacturing Company.

Leonard Ricard, General Motors Corporation.

William Sandras, Hewlett-Packard.

Barbara Lingle dedicated extra effort in typing the manuscript into a presentable format, much more than a nine-to-five contribution.

My wife, Kay, has over the past few years been very patient with my absences from home and has regularly taken care of all the problems that would distract me from the work.

After all this assistance, there are still errors in the book, some by oversight and some because we are venturing into a path of development about which much has yet to be learned.

Much of the inspiration for writing this came from the Repetitive Manufacturing Group of the American Production and Inventory Control Society, a small group dedicated to spreading the knowledge of stockless production in the United States and to advancing the practice of it wherever repetitive manufacturing may apply. The meetings and workshops of this group have contributed much toward understanding the problems of industry and developing ideas for resolving them in the United States.

The work has been sponsored by the Educational and Research Foundation of the American Production and Inventory Control Society. Both the foundation and the society are dedicated to research and education that will improve the quality of work in industry, primarily by improving production planning and control, and in any other way possible. Their generous support of a study trip to Japan and visits to American companies has made it possible to develop the background for this work. The officers of the foundation and of the parent society itself have been supportive, not only with funding, but also with their time and much encouragement.

The American Production and Inventory Control Society is dedicated to the people who must daily make industry work. Recently, everyone in America has become aware that much of American industry has a productivity problem, and sometimes the efforts of the people who work in industry have been maligned in the process. If you are among those who have to make industry work, this book is really dedicated to you—in the same spirit that the American Production and Inventory Control Society has from its beginning stood for bringing people in industry together to share their problems and their solutions.

Robert W. Hall
Indiana University

Contents

1

Stockless production: Basic concepts

Zero Inventories connotes a level of perfection not ever attainable in a production process. However, the concept of a high level of excellence is important because it stimulates a quest for constant improvement through imaginative attention to both the overall task and to the minute details. That leads to practical actions which break out of previously accepted tracks of thought about production.

One such concept of an ideal total production system is most commonly called just-in-time production, a name which emphasizes producing exactly what is needed and conveying it to where it is needed precisely when required. The misimpression sometimes exists that just-in-time production and ideas associated with it apply only to automotive production or to high-volume manufacturing which is similar to it; but that is not true. Stockless production is a name sometimes used to mean the same thing, usually implying application in a broader context, and stockless production is the name used for the system throughout this book. Whatever it is called, the ideas need to be considered and understood in their entirety because they encompass every aspect of manufacturing management.

The goal of stockless production is to find practical ways to create the effect of an automated industry which will come as close as possi-

ble to this concept of ideal production. It is not an end in itself, because the pure ideal cannot be literally attained. It is a guide to constant improvement, by big steps and small ones, to bring ourselves ever closer to the ideal. It causes us to see problems we never before recognized and to develop better and better techniques to solve them. It constantly promotes the simplest, least costly means for every possible aspect of manufacturing practice. There is no point striving for lesser goals if effort can be spent in equally practical ways to change manufacturing into what it really can be.

Instantaneous conversion of raw material to finished product is, of course, not literally possible. The benefit comes from studying very carefully what we would want such a system to do and then seeking ways to transform the current production system into something approaching as closely as possible what we want. So what should such a system do?

1. Produce products the customers want.
2. Produce products only at the rate customers want them.
3. Produce with perfect quality.
4. Produce instantly—zero unnecessary lead time.
5. Produce with no waste of labor, material, or equipment—every move with a purpose so there is zero idle inventory.
6. Produce by methods which allow for the development of people.

The adoption of this philosophy removes many barriers in thinking, and that, in turn, leads to many new techniques for production. But stockless production is more than a set of established techniques to be installed. It is a fundamental way of thinking to transform overall manufacturing in the simplest way possible and to generate new and original techniques for doing so.

FLEXIBLE AUTOMATION: SHORT SETUP TIMES

The ability to produce one unit of whatever is wanted by the customer at any time requires flexible automation. Flexibility means that plants should be capable of switching very quickly from one product to another, or from one part to another, within the capability for which equipment is designed or modified. This means automation of setup processes as much as automation of running operations. The purpose is reduction of setup times. If possible, a piece of equipment should be able to change from running one part to another almost instantly.

Flexible setups can be accomplished through programmable equipment if the physical process of changing tools and attachments requires little time. Since programmable equipment costs money, a manufacturer should try to reduce the setup times as much as possible on the existing equipment, using simple methods and as little investment as possible. Any manufacturer is better off with short setup times than long ones if shorter setup times can be realized at little cost. Many kinds of equipment can be modified to attain very short, flexible setups at low cost.

MARKET-PACED RUN RATES + FLEXIBILITY = SMALL LOT SIZES

Producing only at the market rate of demand and without producing inventory means making parts and products in small quantities—building today only what is needed, nothing more. If parts are fabricated and flowed together into assembly, and if only one product at a time is demanded, then only one set of parts for one unit is demanded at a time. The ideal lot size is therefore one.

This is one of the most difficult parts of stockless production to accept. It is easy to become distracted by all the existing limitations that prevent a lot size of one from being realized and easy to think that larger lot sizes must be better. The existence of large lot sizes means only that we have not yet mastered the technology of production and transportation to make just one unit at a time when we want one.

If production needs to run only at market rate, then high-speed automation is not necessary. There is no reason to run a month of production in five minutes only to have the machine sit for another month. In such an instance, everyone sees that production does not have to be high speed unless the kinematics of the process require it, and even in that case a different process might be possible. The economics of lot sizing are not permanent economics, and the objective of management is to change the economics of running rates, setup times and transportation. Stockless production means looking for ways to change parameters, not accept them.

DEVELOPING THE MANUFACTURING PROCESS TO DO IT RIGHT THE FIRST TIME: PERFECT QUALITY

To have a fast flow of parts through production in small lot sizes, quality coming from each operation must be excellent:

No rework.

No substitutions of tools or materials.

No overproduction to allow for normal scrap.

No scrap for adjustments or "run-ins."

No damage in transit and handling.

If this can be done, then screening out defects by inspection after production is complete is unnecessary. In order to do it right the first time, the process must be clearly understood. The emphasis goes into attaining process control so that any defect seen can be given an assignable cause and the cause eliminated. This is a process of constantly working to make the production process as error-free as possible. Through cutting inventories and lead times we pressure ourselves to reduce error rates to minuscule amounts, not by expecting people to become perfect but by making the equipment and procedures themselves as fail-safe as possible when operated by ordinary mortals.

This approach to quality extends beyond the physical processes themselves. It is an ongoing effort to reduce the variance (uncontrolled occurrences) in every part of manufacturing: planning, design, and execution—until waste and errors become virtually nonexistent. Even better than correcting the causes of defects which have already occurred is the practice of anticipating the causes and correcting them *before* they occur.

REDUCING PRODUCTION OPERATIONS TO THE MINIMUM NECESSARY

Sometimes complex equipment and systems are developed with insufficient basic thought on how to do what is needed at the right time without error. The objective is to eliminate unnecessary activity and complexity. Move material as directly as possible from its existing state to its finished state so that every move adds value to it. This is the basic precept of industrial engineering, but it can easily be obscured by traditions that violate it. The ideal of stockless production drives people to examine operations in detail for the purposes of

Eliminating waste of *time*. Nothing sits.

Eliminating waste of *energy*. Operate equipment only for a productive purpose.

Eliminating waste of *material*. Convert all of it to product.

Eliminating waste from *errors*. No rework.

There is no point in moving material from one point to another just so it can sit in a different place. Double-handling costs money for transport and checking. Also there is no point in executing any operation in production if it can be avoided, and that is true whether it is performed by simple hand tools or programmed equipment.

Shigeo Shingo gives an excellent example of eliminating an unnecessary operation.[1] Several years ago he studied the best ways to remove the flashing from die-cast parts. He tried to find a quick, direct way to remove the excess metal without destroying critical dimen-

[1] Shigeo Shingo, *Study of Toyota Production System* (Tokyo: Japan Management Association, 1981), p. 11.

sions or appearance. However, he saw Daimler Benz in West Germany using a process of vacuum die-casting which eliminated the formation of flashing. (Flashing is formed by metal entering the gaps left between die halves for air to escape. However, if a vacuum is drawn on the mold cavity just prior to the shot, the die halves can fit so closely that flashing is not formed.)

This is not particularly novel. Industry all over the world is in a constant search for operations that can be simplified or eliminated, and there are thousands of examples daily, some dramatic and some trivial. The major reason there are not many more of them is that the objectives of manufacturing are not always focused on accomplishing more of them. Stockless production gives careful and unswerving attention to simplifying operations.

Examples: A high-bay automated storage facility left almost empty when it was discovered that the plant could run without a stockroom. A monorail conveyor left idle when it was discovered that modified carts conveyed material with less throughput time. A numerically controlled, 12-station transfer machine eliminated when it was found too inflexible to modify for fast setup times.

Stockless production begins with the design of the product. A product that functions well must not only contain the features wanted by the customer, but it must also be produced well. The idea is to plan for production during design so as to make each product enter production as far down the learning curve as possible.

Stockless production extends to the design and modification of equipment. Simpler equipment is simpler to modify and maintain. By the laws of reliability, complex equipment is more likely to have downtime from failures that are difficult to diagnose. That does not mean that equipment should not use advanced technology, but only that it should not be more complex than is necessary. A robot with a full range of positional capabilities is not needed if the total task consists of two-dimensional repositioning. The logic of this is obvious, but it is often violated if engineers or managers are uncertain what really needs to be done and use equipment with capabilities not really needed.

Stockless production guides a balanced development of equipment and operations. It concentrates engineering on what is most important instead of on efforts that may turn out to be wasted. The principle is to first improve the operation until one is certain what is necessary, then improve the equipment and facilities.

PREVENTIVE MAINTENANCE TO IMPROVE PRODUCTION CAPABILITY

By stockless production concepts, scheduled equipment time includes time for checking and maintenance. Unless equipment must be run around the clock, as in extrusion operations, a little time should be scheduled each day for checking and maintenance to be sure that it remains capable of high-quality production at all times. There are other reasons for having a little scheduled downtime each day:

Equipment should be in condition to produce whatever is needed when it is needed—not possible if running on a maximum output schedule.

Equipment operating in a tightly balanced flow of material can never catch up if there is no slack time provided for it.

Most of the argument against preventive maintenance is that it costs more for regular downtime and routine refurbishment than would be true if the equipment were run until repair is unavoidable. Such a policy risks losing process control because equipment capability degrades. Repairs often take place under such time pressure that they are not done well, thus starting a cycle of adjust and tinker. The real reason for preventive maintenance is to preserve and enhance the capability of the equipment. That decreases the likelihood of equipment as a cause of unexplained defects in production. In fact, the performance of well-maintained equipment is likely to improve over time, for several reasons:

Operators and maintenance personnel are more familiar with equipment and its problems if they service it frequently.

There is less chance of a process drifting out of control if machine and tool records are tended religiously through preventive maintenance.

Quality, flexibility, and reliability of equipment is almost always improved.

Preventive maintenance is necessary for continuous, long-term improvement of the quality of the production process. Thinking of it in terms of cost trade-off makes the unconscious assumption that long-term improvement is not a major factor to be considered.

STOCKLESS PRODUCTION: ACCOMPLISHED BY PEOPLE

Any system is at bottom a human system. Even if progress has advanced so far that material flows completely through an industry untouched by human hands, the control and further advance of production methods are done by people. Even if all the machinery is programmed, it has to know how to produce the desired result. Automation may displace manual labor, but it cannot displace the human element and be successful.

Conversely, part of stockless production is that no person should be consigned to perform a task which a machine can do economically. In order for the machines to perform well, people have to do an excellent job of programming them, debugging them, maintaining them, and improving the way in which they perform their task. Otherwise, the system will degrade.

Stockless production does not start fully automated. People have to work through what a system is supposed to do, develop it, and refine it. In human terms, that means that no worker should be regarded as a mule useful only for manual activity, but as a worker in a production laboratory, always working to make improvements. That simple change in intent, if it is put into practice through performance measurement and allocation of responsibilities, changes the atmosphere of a company.

A system as tightly linked as a stockless production industry requires skilled teamwork developed over a period of time. The system itself must stimulate this. On the plant floor, the system must present obvious signals of what is to be done. It is not possible to have the fast, flexible material flow and anticipation of potential quality problems without it. The same is true of the system established between suppliers and customers.

Stockless production is a system that requires very close team development and control. Members of each team should have immediate awareness of what needs to be done, which is not possible if the communication system is sluggish and has long lag times.

Consider an example, one developed by evolution through stockless production, the Kamigo Engine Plants of Toyota Motor Company. The plants are almost completely automated. In one of them, 161 people, divided into two shifts, complete 1,500 engines per day. In the center of this plant hangs a signal board large enough to be seen by everyone. This board shows the planned production rate of engines, how many engines have been completed so far today against the 1,500-per-day schedule, and which points in the plant are operating at that rate and which are stopped. By looking at the board, workers can tell whether they are on schedule or whether they need to catch up or slow down. They can also see where the trouble spots are. Feeding material into an operation which is down only jams the system.

In addition, many of the individual machines also have status lights on them visible from a distance. By knowing the signals of the lights, workers can tell at a glance what machine needs attention and something of what is required: repair of a malfunction, a tool change, addition of more material, a quality check, and so forth. By this system, workers can service many machines as a team. One worker does not stand and watch a machine just to be sure that nothing goes wrong—a deadly boring assignment. They have to work together, and they have to think.

This is only one part of the visibility systems that are a part of the stockless production philosophy. Throughout this book, many examples of such systems are presented. People must develop them, and people must develop themselves to work as a team using them. The physical nature of the various systems of visibility has great variety; each one must be developed to suit the particulars of the department, plant, or industry of which it becomes a part. The constant is the need for people to function as a team.

Teamwork does not overcome the problems of personal animosity, warring factions, or debilitating habits that seem to be a part of the human condition; so people who work in teams may not live in complete harmony, but a team must be capable of functioning as such. It takes experience and development to do it.

Stockless production is learned by doing it. This book presents some of the fundamentals, but no book can make sufficient preparation. It is like learning to swim. The art of innovation in reaching for the ultimate production system must be practiced. It cannot be fully understood vicariously.

Developing ourselves for it is a major challenge in stockless production, perhaps *the* major challenge. It requires finding the ways that people can perform together for the various kinds of activity required. This cannot be done by adopting the forms of teamwork (such as quality control circles) without putting what is to be done together with how to do it. There is a heavy element of on-the-job development required in order to physically see problems and know what to do.

Any number of major issues in developing people can potentially arise with stockless production. Job security and improvement-oriented performance measurement are but two of the most obvious. They have to be handled well as they arise if stockless production is to be successful.

DEVELOPING AN ENTIRE INDUSTRY AS A SINGLE FACTORY

From the point of view of a physical process, a supplier is but a remotely located work center of the customer company. That work center happens to be shared with other companies. The objective is to design the physical process so that the supplier's last operation hands off material to the customer's first operation with as little wasted activity as possible, just as if the operations were physically side-by-side. In that way, the concept of stockless production is extended to an entire industry, a physical model of company relationships—not a business model, and that implies several things:[2]

1. Try to reduce the randomness in material flow. Parts usually leave the last operation of a supplier one at a time. They usually enter the first operation of the user one at a time. Why gather them into groups and redistribute them one at a time if that can be avoided?
2. Therefore, try to make economical the transport of materials in small lot sizes with a high frequency, just as between work centers in a plant.
3. If this can be done, the inventory levels and lead times between supplier and customer can be cut.
4. All this leads to joint activity between customer and supplier to improve the production operations at both locations, and this is the real objective.

One of the most important improvements is likely to be in quality. If perfect quality is important within a plant, it is also important between plants. However, it is not obtained without great effort. The routine communication between customer and supplier should provide feedback and assistance to be sure that all operations are performed correctly the first time, just as is true of the quality programs within plants. A product is a product of an industry, not just a single company.

The development of this system of direct hand off of material throughout an industry suggests many aspects of supplier relationships which do not now prevail in much of industry. Supplier-customer relationships need to extend over a period of time, and both parties need to trust each other with more details of their operations than is often true. This means more faith that each will not try to invade the other's business (usually expressed by customer companies who decide to "pull business in house").

A supplier in an industry does not serve a market in the same sense as does a manufacturer who sells a product to the customer who will *use* that product. The supplier serves the real market indirectly through the companies that buy its output for use in end products for either consumers or companies. Therefore, an industry is a collection of companies and operations that together serve the end market. A gasket performs a vital function in an engine which in turn goes into an automobile which someone then uses. Likewise, an integrated circuit has a vital function only when it is used in a computer that someone uses.

---

[2] This is an extension of the idea of a focused factory, first developed by Wickham Skinner. See "The Focused Factory," *Harvard Business Review*, May–June 1974, p. 113.

If all the companies in an industry recognize that the real competition is for end item users, the nature of the industry as a physical system begins to clarify. It is very important for each company, and for each plant within various companies, to improve according to the principles of stockless production, but at some point a complete industry must improve. This begins by customer companies reviewing how many suppliers they have and what purpose they serve. Is each supplier needed because the customer company lacks space, lacks skilled people, cannot directly manage a large enough production operation, and so forth? Once a firm basis of need for each supplier is recognized, the basis for long-term relationships with them is established. Then companies in an industry can form partnerships to jointly improve operations.

Of course, any given company as a business entity may operate a number of plants that serve different roles in different industries. However, if there is difficulty establishing the production mission of people and equipment at each plant, there is difficulty determining the basis for improvement.

STOCKLESS PRODUCTION: A REVOLUTION FOR BUSINESS PLANNING AND STRATEGY

Stockless production is a way of looking at plants, companies, and industries as physical systems. The pattern of thinking emphasizes what must be done to improve operations to the highest degree possible. Therefore, business planning starts from physical concepts and develops a plan. It does not start from a business plan that considers the physical side in a vague way. Because of this, stockless production obviates much of the trade-off thinking that is the basis of the cost side of much business planning.

Many business plans call for improvement compared either with the performance of competitors or with previous performance. An arbitrary goal of a 5 percent cost reduction is an example. That is a 5 percent cost reduction compared with last year. It may be set forth as a goal in hopes of stimulating improvement activity not specifically envisioned.

Stockless production compares performance with a vision of what is ultimately possible. It provides a point of reference to constantly remind us that better is possible if we can find the way. If the physical operations can be developed by stockless production, the financial results follow.

Production is only one of many elements in the overall strategy of running a company. The features of the products, the marketing plan, research and development, cash flow, acquisition of capital, and other equally weighty considerations are intertwined in a good overall business strategy. However, stockless production is not confined to a set of techniques for improving production defined in the narrowest way as material conversion. It is a way to visualize the physical operations of the company from raw material to customer delivery, but there is no aspect of management which it does not touch.

INVENTORY LEVEL AS AN INDICATOR OF TOTAL PRODUCTION EFFICIENCY

One of the primary ways to force attention on the problems that prevent stockless production is by reducing inventory levels. If production flexibility and process control are improved, inventory can remain at low levels.

Much more is involved than merely reducing the *investment* in inventory. Once a company has reduced inventory to, say, 12 days of work-in-process plus raw material, cutting the inventory in half again does not usually represent a great financial impact in the inventory account itself. Much more significant is the impact of all the improvements required in order to accomplish that. Inventory level is therefore a measure of total production efficiency, the financial effects of which appear in many ways due to short lead times, low defect rates, smooth material flows, and high productivity.

The real question is never "How much inventory is needed?" but always "*Why* is inventory needed?" The existence of inventory is evidence of an impediment to stockless production, and impediments should be removed. Some can be easily removed, and some, it appears, are impossible to remove with present knowledge; nevertheless, the objective of inventory management methods is to force frequent review of the details of production operations so as to consistently reexamine the problems that require inventory.

A major milestone in this process is to operate with no active production inventory kept in a stockroom at all, but it does not end there. The real goal is to have no inventory anywhere that is not actively in process. The term *stockless production* was coined from this idea, but it connotes attacking all the problems that require inventory.

That inventory is a symbol of problems is almost universally recognized at times, but not constantly. For example, when the physical inventory is taken in most manufacturing companies, examination of the results generally shows that the inventory contains the residual of everybody's mistakes: transaction errors, double orders to be sure that at least one order arrived, extra safety stock to cover for machine breakdowns, leftovers from quality inspection disputes that were never resolved, and on and on. Writing down the shortages and losses purges the organization of the evidence of past errors. It does nothing to preclude a repetition of those same errors in the future, and the inventory kept because setup times are much longer than necessary is often not recognized as such.

Excess inventory does not exist only in stockrooms. In most plants, almost every operator by instinct keeps a little extra inventory of something on hand just in case everything does not go smoothly. A little safety stock, whether formally recognized or not, exists in many places to cover for difficulties. It takes a great deal of resolve from both managers and workers to stop this and concentrate on what can be done to reduce the variance in production—the unplanned downtime and rework that seem necessary to overcome Murphy's Law in all its versions.

Cutting the level of inventory stimulates action by speeding the flow of material and decreasing the throughput time. To do that requires physical and procedural changes to improve flexibility and process control. That is fundamental. Therefore, management should seek to cut inventory in those areas in which it wishes to concentrate improvement effort—a work center, a department, or a plant where action programs are wanted. One of the most common analogies for this process is the level of water flowing in a stream. Where the stream is deep, the current is slow; and where it is shallow, the current is swift. The objective is to allow the same volume of water to flow downstream with a uniformly swift current and a stream depth as shallow as possible.

Normally, a picture of a boat going down a stream with rocks on the bottom is used to illustrate this point, but the interpretation must be done with care. The objective is to remove the rocks from the stream bed so that it is smooth enough for a boat to traverse in safety when the water level drops low enough for them to be seen. Some of them can be seen below the surface of the water, and some of them are unseen until they appear above water. Some of the rocks may be easy to remove and some may be very difficult. If one is interested only in safe navigation, keeping the water level above the rocks is a good way to accomplish that. It is somewhat difficult to keep in mind that the objective is to lower the water level and remove the rocks.

The illustration in Exhibit 1–1 contrasts two different channels of flow from a dam. The upper and lower water levels between the two reservoirs are the same, so the volume of water flow is the same in both cases. However, the rocks trap pools of slow-moving water on the way to the lower reservoir until they are cleared. Then the water runs in a swift, smooth flow. This analogy is a better one to illustrate reducing the water level. Flow analogies are used many times in discussing stockless production.

Another analogy which illustrates the effect of inventory is that of a truck convoy. If the interval is 500 yards between each of 20 trucks in a convoy and the lead truck is driven at a steady 50 miles per hour, the intervals between trucks remain far from constant. At such a distance, a driver may be within a hundred yards of the truck ahead before he

EXHIBIT 1–1 Analogy of water flow to material flow in production

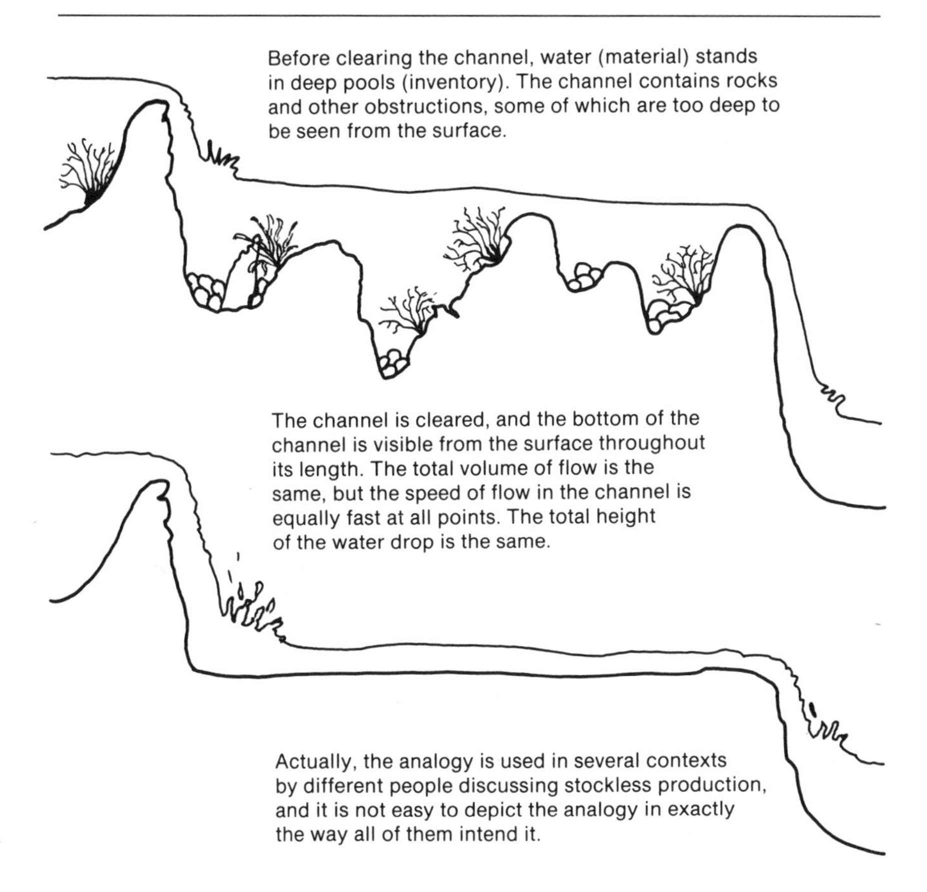

notices that he is too close and slows. Conversely, the truck ahead may completely disappear around a curve when the interval opens beyond 500 yards, so the driver behind races to catch up. The net effect is a constant telescoping of the intervals between trucks until the 20th one is either completely stopped or has the gas pedal mashed to the floor.

Consider the same convoy of 20 trucks if the interval between is reduced to 100 feet and the lead truck is set on 50 miles per hour. Now alertness is required for survival. Every driver becomes greatly interested in whether the driver behind has brakes that work and whether the driver ahead has brake lights that work. Most likely, everyone involved will give great attention to the condition of all the trucks and all the drivers. Everything should be in perfect condition.

Furthermore, the drivers may start to think of better ways to control the entire convoy. They may want to see a signal if the lead driver is even so much as thinking of changing speed, so they begin to create a better signaling system to indicate what is happening to each vehicle, and that includes better signaling systems for mechanical problems of each truck. In the same way, reducing inventory forces everyone to pay very close attention to the problems that required the inventory and to think about managing with a set of signals from the process itself (trucks), not from the interval between them (inventory.)

Most of us have misgivings when we first begin to consider stockless production because it takes time to develop a complete framework of thought about it. The key point in overcoming these is to keep in mind that the idea is to make long-term changes in how production is accomplished until the best imaginable system can be made to work. Here are a few of the common misgivings:

It will take enormous investment to reduce set up times and increase flexibility. For many common types of equipment, reduction of setup times can take place with relatively inexpensive modification. As setup times decrease, approximately the same amount of production time spent on setups will allow for many more setups per day.

Material handling costs will increase. Stockless production means changing layouts, reducing the distances between operations, using different containers, eliminating double handling, transporting more mixed loads of material and other physical changes designed to reduce material handling cost in the long run.

Employee opposition will be too strong. Whether unionized or not, workers will certainly be apprehensive if they do not understand it. So will many members of management. Unless these fears are overcome, it will not be possible to have the leadership necessary at the plant floor level. However, if management pursues the kind of leadership needed for stockless production, that in itself will overcome some of the most acrimonious issues between management and labor.

We will have to change our entire culture to make it work. It is not necessary to change basic human nature. Stockless production is a managerial revolution, not a cultural one. As we shall see, many of the ideas embedded in it have been used piecemeal in plants in many parts of the world.

We have too many problems to operate that way. Stockless production is an approach for overcoming problems. It is much more than a control system for streamlined flow of material.

These misgivings are but a sample of many, and most manufacturing managers have an understandable degree of uneasiness about abandoning the patterns of thought that have guided them for years. One of the most pervasive of these is trade-off thinking typified by the use of economic order quantities (EOQ), or economic run quantities. Recall that the basic EOQ equation is:

$$Q = \sqrt{\frac{2 \cdot S \cdot D}{i \cdot C}}$$

The S in the numerator is setup cost, or order cost. In use, the formula assumes that this cost will be a constant, but it does not have to be. In fact, if it were reduced to zero, the EOQ would be zero. The intent of stockless production is to shift setup costs into fixed cost without increasing fixed cost or, in other words, to lose no necessary production time and incur no extra cost due to setup. If that is done, the setup cost is no longer something to consider in the old way.

Trade-off thinking embedded into cost systems and performance measures is of more importance. The EOQ is only a lot-sizing technique, and it has had less impact than the cost systems, particularly as applied to efficiency measures. Most of these are variations of earned hours divided by actual hours worked, and much energy goes into developing them and appraising their results.

The important issue is what the emphasis on efficiency causes the personnel to do. To maximize the standard hours earned, they want to keep the equipment running as much as possible. Even if setup hours are part of the standard, no parts are made while the equipment is being set up. The way to maximize standard hours is to avoid setups—run as long as possible. By inference, this puts an extremely high cost on setup time. It encourages supervisors to produce as much as possible, even if it is not needed. It also encourages them to do "gravy jobs" first, that is, to run as much as possible of the parts in which there is the highest ratio of potential standard hours to earn versus actual hours worked. It discourages setups. In practice, it does not generally encourage supervisors to put much effort into refining the skills and practices for setups.

Just as important, both EOQ and efficiency measures are generally too narrow in scope. They put very little emphasis on improvement of the system because they assume that supervisors are responsible only for operating efficiently by methods that are essentially fixed over short periods of time—weeks or months. Often the measures of performance used at higher levels of a company are little better. Operations become managed by cycling through a series of campaigns to improve according to various short-term goals: inventory levels, cash flow, quarterly profit, and so on. These tend to work against each other and provide little direction for long-term improvement.

By contrast, the *trend* in inventory level, as measured by days on hand, is a good measure of overall progress, but it is not sufficient. After all, a company could have practically no inventory and be in extremely bad condition. However, if output is on target, defect rates are down, customers get what they want on time and costs are low, then inventory level is something of an overall measure of the physical changes that have been made to do this. Most important is to measure people and results by trends of improvement.

Reducing the inventory level by which everyone operates forces them to think about how to correct problems that could be covered by safety stock. Some of the problems can be easily corrected, and some take a long time. The system should cause people to give daily attention to improvement, no matter how automated the production becomes.

BROAD SCOPE OF APPLICATION

Stockless production in a general sense applies to any kind of production. Improving process capabilities, reducing setup times, and similar activities are surely not harmful to any manufacturer. However, most of the detailed discussion of this book concerns repetitive manufacturing, one of several kinds of manufacturing classified in Exhibit 1–2, which is a modification of a classification developed by Harvard Business School faculty.

The definitions of the different kinds of production are those used in this book. Manufacturing definitions are far from standard, and, likewise, it is not possible to cleanly assign some kinds of manufacturing to a category. Furthermore, many manufacturers have several kinds of production under one roof.

From the discussion so far, it is already evident that stockless production promotes repetitive manufacturing. Over time, many manufacturers have transformed from a job shop operation to repetitive as their volume increased. Henry Ford made the transition beginning with his first assembly line. This conversion means reducing the inventory between operations, transferring workpieces in a flow, and decreasing the throughput time of material. Stockless production has the same objective, only more so, and one way to think of it is as an attempt to make an entire industry perform as if it were all part of a single assembly line.

There is, however, a great deal to be learned from stockless production which can be applied to a job shop. Stockless production has not been applied perfectly to any industry, so it is a question of how smoothly a job shop can be expected to perform. Many manufacturers who operate production by job shop methods may not need to do so. It is worth a review to determine if the production in a job shop is potentially repetitive. Even if that review turns out negative, most of the principles of stockless production apply except for refined smoothing and leveling of material flow.

EXHIBIT 1–2 Classification of manufacturing by physical processes and by managerial methods

	Type of conversion process	
	Discrete units (fabrication and assembly)	Nondiscrete units (physiochemical conversion processes)
Repetitive manufacturing (discrete) or continuous manufacturing (process)	Autos Appliances Box wrenches	Oil refineries Steel rolling mills Fertilizer plants
Job shop manufacturing	Custom machine shop Cabinet making Print shop	Fermentation processes Paint mixing Kitchen
Project manufacturing	Construction Shipbuilding	Specialty chemicals

Repetitive manufacturing: Production of discrete units planned and executed by a schedule. Material moves in a flow. (If the product is a gas, liquid, or powder, this is usually considered continuous production, as in the extreme case of an oil refinery.)

Job shop manufacturing: Production of discrete units planned and executed in irregular-sized lots. Material moves in integral lots through production.

Project manufacturing. Producing units of such size and complexity that each one generally requires planning and control by its own project organization.

Repetitive manufacturing	*Job shop manufacturing*
Plan and schedule by rate.	Plan and control by lot sizes.
Fixed flow path of material.	May have alternate routings.
Equipment dedicated to a range of tasks which allow flow.	Acceptance of a broad range of tasks for equipment.

There are valid reasons which require job shop methods. Some of them:

Many unique custom orders which require custom engineering.

Operations performed are "on the edge of the state of the art," which means that yields are poor and sometimes unpredictable.

Quality control requires inspection or holding in lot sizes. Example: pharmaceuticals.

Small-volume orders arrive at randomly spaced, irregular times.

However, it is easy to think that a valid condition exists that demands a job shop operation when that may not be true. An imaginative review of the principles of stockless production may show that there is really a way, or that it applies to a substantive part of the operation, if not all. Stockless production may apply in some degree to a combination of types of manufacturing, though it is most fully developed when perfecting production that is already repetitive, or when converting from job shop to repetitive. As with almost anything that is difficult, it is easy to disclaim applicability.

Stockless production requires a positive attitude, a determination to seek out problems and solve them, and nonacceptance of the belief that barriers to a smooth flow of production are permanent. No company which has embarked on it has attained all they wished to do, but they are much better off than if they had decided not to try it at all.

APPENDIX: GLOSSARY OF TERMS

Attachment An item such as a die, tool, adaptor, or extension that may be exchanged at the time of a setup. By contrast, the machine itself remains unchanged and usually provides frame, power, and master control.

Automation The substitution of machine work for human physical and mental work, or the use of machines for work not otherwise accomplishable, entailing a less continuous interaction with workers than previous equipment used for similar tasks.

Backflush The deduction from inventory of the component parts used in an assembly or subassembly by exploding the bill of materials by the production count of assemblies produced.

Balancing operations In repetitive stockless production, trying to match actual output cycle times of all operations to the cycle time of use for parts as required by final assembly, and eventually as required by the market.

Broadcast system A sequence of specific units to be assembled and completed at a given rate. This sequence is communicated to supply and assembly activities to perform operations and position material so that it merges with the correct assembled unit.

Cause-and-effect diagram (Also called *fishbone diagram* or *Ishikawa diagram.*) A precise statement of a problem or phenomenon with a branching diagram leading from the statement to the known potential causes.

Continuous manufacturing Production of nondiscrete material without interruption as in extrusion, distillation, rolling or calendering operations, and so on.

Control chart A statistical device usually used for the study and control of repetitive processes. It is designed to reveal the randomness or trend of deviations from a mean or control value, usually by plotting these.

Cycle time The time between completion of two discrete units of production. For example, the cycle time of motors assembled at a rate of 120 per hour would be 30 seconds, or one every half minute. (*Cycle time* among materials managers often refers to the amount of time from when material enters a production facility until it exits. In this book it is referred to as *throughput time*. Cycle time is used in the industrial engineering sense, not the materials sense.)

External setup time Elements of a setup procedure performed while the process is in production (the machine is running).

Failsafe work methods Methods of performing operations so that incorrect actions cannot be completed. Examples: A part without holes in the proper place locks in a jig. Computer systems reject invalid numbers or require double entry of transaction quantities outside normal range. (Also called *foolproof methods*, but that term is sometimes offensive to workers.)

Flow rate Running rate; the inverse of cycle time; for example, 120 units per hour, or 2 per minute.

Immediate feedback The principle of inspecting production as quickly as possible after a part is made and providing information on deviations from desirable output standards so that correction is immediate. Extends to automatic shutdown of a process in case of a defect or a drifting out of desired control range. Loosely applied to feedback on any problem for immediate action.

Inbound stockpoint A defined location next to the place of use on a production flow to which materials are brought as needed and from which material is taken for immediate use. Used with a pull system of material control.

Inefficiency stock Stock held because operations are inefficient for any reason—long setup times, machine problems, unskilled operator, and so forth. (See safety stock.)

Internal setup time Elements of a setup procedure performed while the process is *not* in production (the machine is *not* running).

Inventory level In stockless production, this particularly applies to work-in-process inventory levels, sometimes called *depth of process*.

Job shop manufacturing Production of discrete units planned and executed in irregular-sized lots. Material moves in integral lots through production.

Just in time This phrase is used in two ways:

1. Narrow sense: Just in time refers to the movement or transport of material so as to have *only* the necessary material at the necessary place at the necessary time.
2. Broad sense: Just in time refers to all the activities of manufacturing which make the just-in-time movement of material possible—i.e., stockless production.

Lead time A span of time required to perform an activity. It includes time for order preparation, queuing, receiving, inspection, transport, and so forth. Not the same as throughput time.

Level schedule A schedule such that the use of all parts in materials is as evenly distributed over time as possible. *Level* means level use of all parts for a given period of production.

Line balancing Reassigning and redesigning work done on an assembly line to make work cycle times at all stations approximately equal.

Linear production Actual production to a level schedule, so that plotting actual output versus planned output is straight, even when plotted for a short segment of time.

Machine For purposes of setup, those basic elements of equipment that do not change during the setup.

Move card or move signal A card or other signal indicating that a specific number of units of a particular part number are to be taken from a source (usually outbound stockpoint) and taken to a point of use (usually inbound stockpoint).

Outbound stockpoint Designated locations near the point of production on a plant floor to which produced material is taken to await pickup by a pull system.

Pipeline stock The amount of stock held between the point where it is made and the point where it is used.

Process capability The basic physical capability of production equipment and procedures to hold dimensions and other characteristics of products (such as electrical resistance) within acceptable bounds *for the process itself*. Not the same as tolerances or specifications required of the produced *units* themselves.

Process control The ability to maintain a process within a given range of capability, by feedback and correction, for example.

Process manufacturing Production which adds value by mixing, separating, forming, or chemical reactions. It may be done in either batch or continuous mode.

Production card or signal A card or other method for creating notification that a specific number of units should be made for use or to replace some removed from pipeline stock in a pull system.

Production network The complete set of all work centers and process and inventory points, including all suppliers, from the point of completion of the product all the way back to the raw material for all components.

Production rate The number of items completed per unit of time; same as flow rate.

Production time The time for a production card to make a complete circuit through a producing work center, allowing for all the operations *necessary* for the production: setups, inspection, material placement, and so on.

Project production Production in which each unit or small number of units is managed by a project team created especially for that purpose.

Pull system The production of items only as demanded for use or to replace those taken for use.

Push system The production of items at times required by a given schedule planned in advance.

Quality control circle An organization of 5–12 people who normally work as a unit for the purpose of seeking and overcoming problems concerning the quality of items produced, process capability, or process control.

Queue time The amount of time a job waits before processing at a work center. Usually used in context of job shop production.

Repetitive manufacturing Production of discrete units, planned and executed by a schedule, or by rate. Material moves in a flow.

Rhythm quality control Quality control actions such as inspections or plotting control charts and process checks done in a regular pattern integrated into the daily routine of production.

Safety stock A stock of items known to be of good quality; kept in reserve and used to replace occasional items found defective or damaged in further processing. Part of the reason for use is to keep tabs on occasional defects.

Setup time The time required for a specific machine, assembly line, or work center to convert from production of one specific item to another. The time is the elapsed time from production of the last unit of Item B until production of the first good unit of Item A. Downtime of the production process for this purpose. (See also internal setup time and external setup time.)

Signaling system A set of signals to perform specific actions in production at specific times or in response to need for material.

Small-group improvement activities A name for organizing small groups of 5–12 people who normally work as a unit for the purpose of seeking and overcoming all kinds of production problems associated with their work environment. An extension of the mission of the quality control circle.

Stockpoint A designated location in an active area of operation into which material is placed and from which it is taken. Not a stockroom isolated from activity. A way of organizing active material for easy flow through production.

Throughput time The elapsed time from when material starts into a process until it finishes—through a work center, through a plant, or through an entire production network.

Tools Items (wrenches, for example) used during a setup procedure but which are not attached to the machine during setup.

Time bucket (or bucket) A length of time summarized in one columnar display in a production plan or material requirements plan; usually in days, weeks, months, or quarters.

Valve inventory A stockpoint which is too large to be located next to the point of use of material and from which material is drawn by a pull system. Often a plant receiving area.

Variance (production) The deviation of actual occurrences from planned times or expected procedures for all kinds of reasons stemming from both known and unknown causes. Concept related to variance in statistics.

Variance (statistics) The arithmetic mean of the squared deviations of individual items from their arithmetic mean (or occasionally from some other central tendency or from a forecast value.)

Waiting time The *necessary* throughput time from when material is taken from point of production until it is used. As used in stockless production, it includes move time or transport time, as well as idle time spent waiting to be used; and this differs from the conventional APICS definition of waiting time. In conventional terms, it is more like move time plus queue time.

2

Results of stockless production in practice

Over the past 25 years the Toyota Motor Company assembled ideas from many sources and combined them with their own thinking to create stockless production. The origin of some of the ideas is unknown, but Mr. Ohno of Toyota, who is generally credited as the father of it, says that much of their early thinking was influenced by Henry Ford's concept of integrated assembly. In fact, the compactness of the plants at Toyota City was thought to be an imitation of the system devised for converting iron ore directly into automobiles all within the Ford River Rouge Plant. One story has it that the idea of the production control system originated from watching shoppers in a Los Angeles supermarket.

Kanban is the corporate name of the version of the system pioneered by Toyota and companies of the Toyota group. It is a Japanese word meaning card, a definition which conveys very little of the total scope of the system. Later, Toyota sometimes referred to it as a "just-in-time" system, but that implies only that transfers of material are made just prior to being needed; there is no suggestion that many improvements in all phases of production are necessary to do that. (The strict definition of *just-in-time* is to have *only* the necessary part at the necessary place at the necessary time, or, as an American worker would say, "Don't have anything you aren't working on.")

The major ideas of stockless production were put into place in the years 1960 to 1972 at Toyota. As Toyota's competitors in the auto industry began to see how formidable it was, they began to work on their own versions of the system—using different corporate names for them, of course. However, Kanban has become something of a generic term for the total system.

As the idea of stockless production spread beyond the auto industry, companies adopted different names for it. Many of them, possibly the majority, refer to it as a *total production system,* or by a corporate acronym which comes out to the equivalent. However, even then, they may refer to various subparts of it as *total quality control,* for example. Neither stockless production, nor any other name for it, properly represents how it permeates every aspect of manufacturing.

In 1973, the Arab oil embargo struck a very deep blow to Japanese industry. At that time, Toyota saw that it was in their interest to increase the pace at which their suppliers adopted the system. They also saw it as a national duty to instruct many other companies in the system so as to improve the capability of Japanese industry to retain a competitive position in world market during a very tough period. The system is now spreading from the Japanese auto industry to many other industries. Since about 1980, competitors of Japan throughout the world have seen that they should find some way to improve production operations in order to become more competitive. The method used by Toyota is not the only approach to stockless production, and other companies are working out many variations, but the basic methods are much the same.

Toyota instructed their suppliers well, and by the late 1970s the family of Toyota suppliers was becoming a tightly integrated system of production. In the auto industry, Toyota's competitors are busily working to match or better this accomplishment.

The system began spreading outside the auto industry in the late 1970s. Different companies are in different phases of adopting it. Some have been able to put almost all the ideas into practice, sometimes with modification. Others struggle with implementing only a few of the concepts, and some only cut the inventory a bit, then say they are using the system. Most of them started into it because they feared for their survival.

Not all the stories are successful ones. Stockless production is not a natural extension of Japanese culture in general, or of Japanese industrial culture in particular. All the companies that have given stockless production more than lip service have experienced struggle and frustration; and even if they do succeed in greatly improving production, the reward may not be great. There are other factors in successfully running a business besides the effectiveness of the production function.

On the following pages are presented tables showing what has happened (Exhibits 2–1 through 2–11). Bear in mind that the degree of improvement shown for any company depends on how inefficient it was when the program began, as well as on how far it may have progressed. The amount of imagination required to make improvement in a plant that is already well run may be much greater than for one so confused that its operations are near catastrophe. The improvement figures taken from Japanese plants are nearly all taken from a base of performance that would be acceptable in any part of the world. Results are obtained through hard work by dedicated people. There is no magic.

No one really knows how many companies are engaged in what this book calls stockless production. Only 5 or 10 in Japan are regarded (by Toyota) as being Class A in the practice of stockless production. Perhaps another 30 or 40 companies have attained substantial benefit from performance regarded as Class B. Hundreds more are somewhere in progress toward this goal. There are several failures, but most have attained some benefit, and some much more than others. Exhibit 2–1 shows the results of several Class B companies.

EXHIBIT 2–1 Results of stockless production programs in Japan

Company	*Duration of program*	*Inventory reduction (percent of original value)*	*Throughput time reduction (percent of original value)*	*Labor productivity (percent increase)*
A	3 years	45%	40%	50%
B	3 years	16	20	80
C	4 years	30	25	60
D	2 years	20	50	50

This data was collected in late 1981 by Professor Jinichiro Nakane, Waseda University, Tokyo. The companies did not wish to be identified. The figures represent rough management estimates, as the rounded figures suggest. Labor productivity is estimated as sales in yen adjusted for inflation divided by *total* employees.

The superior progress of Company B may be somewhat illusory, since it was thought to be in the worst condition when starting stockless production.

None of these companies had more than two months in raw material plus work-in-process inventory when they began the program. All had some version of quality circles and quality improvement programs in place when they began, so the measurements of improvement have a high base point from which to index improvement.

Improvements are still taking place at each of these companies, and all of them are perhaps Class B at this point. Stockless production is a condition of constant improvement. All the companies anticipate more and more progress on these goals unless overtaken by a business disaster.

EXHIBIT 2–2 Toyota Motor Company comparisons (about 1978)

Table A: Productivity comparisons in hood and fender press plants

	Toyota	*American Plant*	*Swedish Plant*	*West German Plant*
Setup time (hours)	0.2	6.0	4.0	4.0
Setups per day	3.0	1.0	—	0.5
Lot size*	1 day use*	10 days use	1 month use	—
Strokes per hour	500–550	300	—	—

* The lot size for low-volume items (under 1,000 per month) ranged as high as seven days in 1978. By the author's observation in 1982, six setups per shift were not uncommon in presses up to 500 tons. Stockless production shops seldom ran more than three days' use of any part at once even on large presses. High-volume items might be pressed more than once a day.

Looking at the lot sizes and setups per day, one may be tempted to think that each Toyota press handled only 3 parts while the American ones handled 10. That does not appear to be the case, because these figures are averages with some runs longer and some shorter. Also, the strokes per hour of the Toyota presses were higher, on average, in 1978. No reason is known for that other than that the Toyota presses do appear to chunk along in a steady rhythm with little interruption when running. Material feeding is very smooth.

Table B: Automotive assembly plant labor days to assemble one vehicle

	Toyota Takaoka Plant	*Plant A United States*	*Plant B Sweden*	*Plant C West Germany*
Number of employees	4,300	3,800	4,700	9,200
Number of vehicles per day	2,700	1,000	1,000	3,400
Total labor days per vehicle*	1.6	3.8	4.7	2.7

* Note: These numbers are developed by a straight division. They represent total labor days, not direct labor time comparisons as these are often thought of in the United States. Observation of stockless production plants by the author is that there is much less indirect labor in them, although this is difficult to quantify.

The data in these tables came from Y. Suzimori, K. Kusinoki, F. Cho, and S. Uchikawa, Toyota Motor Company, "Toyota Production System and Kanban System—Materialization of Just-In-Time and Respect-For-Human System" (London: Taylor & Francis, Ltd., 1978).

Toyota and the rest of the Japanese auto industry pioneered the assimilation and application of the broad collection of ideas that make up stockless production, so most of the examples of it come from that industry. It is a repetitive manufacturing industry, and stockless production is most clearly explained in that context. However, other industries in Japan, both repetitive and otherwise, have begun taking up

EXHIBIT 2–3 Turnover ratio of working assets* in automotive companies

Year	Toyota	*Company A Japan*	*Company B United States*	*Company C United States*
1960	41	13	7	8
1965	66	13	5	5
1970	63	13	6	6
1980	62	—	—	—

* In the United States the term *working assets* includes cash and accounts receivable as major components in addition to inventory. However, in order to agree with annual report figures and those given orally, the figures in this table must be inventory turnover ratios only. The heading and all the figures through 1970 came from the paper by Suzimori, Kusinoki, Cho, and Uchikawa cited in Exhibit 2–2. The 1980 figure for Toyota Motor Company came from Mr. Cho directly, but the other three companies are not known in order to make an updated 1980 comparison.

Cho also attributes Toyota's inability to further decrease inventory in recent years to a large variety of production. The company manufactures 14 different car lines, and has a large number of product variations required by different regulations in many different countries of destination all over the world.

The inventory figures for auto companies do not include finished goods. In the case of Toyota, finished vehicles were transferred to the Toyota Sales Company as a separate business entity immediately after assembly in 1980. (The two companies are now merged.) American auto companies also make to order and keep no finished vehicles in stock. At most, 15 percent to 20 percent of their inventory figures would be expendable tooling, spare parts, company vehicles, or other items which are not production inventory.

Note that very early in their history of stockless production, Toyota Motor Company had some very impressive inventory figures in comparison with a major Japanese competitor, not just an American competitor. The company's founder, Kiichiro Toyota, liked to run with a lean inventory because he noted that it stimulated his people to think. By 1960, Mr. Ohno and others were beginning to develop some standard ways to do that. However, Toyota has not run away with the Japanese auto industry because of this. Their domestic market share has held steady at 30–33 percent.

The system began with the auto companies, and it has been a major reason for the impact which Japanese auto companies have had in the United States. No one ventures any comparative figures as being exact, but the Japanese industry uses perhaps 60 percent of the labor hours per vehicle as the American. It pays about 60 percent of the American wage rate. The defect rates obtained in cumulative production are much lower. No one wants to state precise numbers, but the amount of in-plant rework is lower in Japan. The finished units are generally rated better in terms of "fits and finishes" quality and in the number of warranty returns in service. The story has had considerable press. (The comparisons in this paragraph have been used in public presentations by the Automotive Industry Action Group, and they come from internal studies by American auto companies.)

EXHIBIT 2–4 Number of annual proposals by workers at Toyota

Year	*Total number of proposals*	*Number of proposals per person*	*Acceptance rate*
1965	9,000	1.0	39%
1970	40,000	2.5	70
1973	247,000	12.2	76
1975	380,000	15.3	83
1976	463,000	—	83
1977	454,000	—	86
1978	528,000	—	86
1979	576,000	—	91
1980	859,000	18.7	94

Data for 1965 through 1974 are from Y. Suzimori, K. Kusinoki, F. Cho, and S. Uchikawa, Toyota Motor Company, "Toyota Production System and Kanban System—Materialization of Just-In-Time and Respect-For-Human System," (London: Taylor & Francis, Ltd., 1978).

Data for 1976 through 1980 are from *Outline of Toyota, 1981,* a publication of Toyota Motor Company. (Parts of this data appear in several Toyota publications with slight discrepancies.)

This illustrates the way in which Toyota built on stockless production to promote quality control circles into much more than the quality problems of the early years. The Toyota system made problems visible, and over time the workers learned to develop improvements themselves. Some of the proposals were trivial, but some were also very substantial; and by concentrating much of the initiative for productivity improvement on plant floor people, Toyota made the most of simple methods of improvement before advancing to more complex engineering. Note that there was a very big emphasis on proposals from workers beginning in 1973, the year the oil crisis jolted them into greater activity.

The work force has been coached over a period of time to participate in making plant improvements. Naturally, from time to time, criticisms of this surface both in Japan and in the West on the belief that *too much* pressure and responsibility is placed on the workers. The argument is almost the opposite of the one often heard in the West in which workers have jobs that are too routine, stultifying, and trivial, so, therefore, they should be enlarged in scope. More than almost any other company, Toyota has tried to obtain the most from modifying existing equipment and using it in a very smooth flow of production before adopting more technically advanced equipment, and to do this they have relied on what they sometimes call *peopleware.* Other companies in stockless production, even in Japan, do not emphasize this quite so much.

the concepts and applying them to their own circumstances. The example of the Sumitomo wire plant (Exhibit 2–10) was included to show that people in many industries are beginning to apply stockless production even when the conditions for repetitive manufacturing are expected to remain less than perfect.

Also, the examples come from the companies generally considered to be in the upper end of manufacturing practice. However, even the best companies can become better, and the development of stockless production is far from finished. Whenever one visits a stockless production plant, one almost always hears the question, "What step do we take next?"

Many American companies are now experimenting with stockless production in some way. These projects are so new that only preliminary results can now be presented. All the American projects will take time, for the one common denominator of all of them is the tentativeness with which Americans are approaching this form of thinking.

EXHIBIT 2–5 Evolution of Stockless Production at Tachikawa Spring Company

The graph on the following page is difficult to interpret because it is cluttered, but it is worth the effort for the patient. It is actually a simplification of a graph presented by executives of Tachikawa Spring Company during a briefing for a visit on July 6, 1982. In one page it summarizes their development of a production methodology over a period of years. Tachikawa Spring is a supplier of automobile seats to Nissan, and their instruction came from several sources besides Toyota. There is no point in their history at which they embarked on a clear-cut conversion to stockless production; they evolved toward it through a series of programs.

Block Ⓐ: This presents five indexes of different kinds of improvement measures. The first four are based at 1.0 in early 1973, and the fifth one at 1.0 in mid-1975. The values of each of the indexes are shown at three points in time, but only the first one and the fifth one are plotted. Note that

Added value = (Sales) − (Value of purchased materials and services)

The reduction in active part numbers came about through a deliberate campaign to redesign products so as to improve the commonality of parts.

Block Ⓑ: The tracking of labor productivity measured in this way is more normal for a Japanese company. Note that they measure productivity by accounting for *all* employees: direct, indirect, and executives.

Block Ⓒ: The number of press setups per month is considered to be an average figure for all presses in use, that is, the number of setups *per press* per month. The time period during which space was reduced by 40 percent was unclear, but mostly took place over approximately a five-year period from 1977 to 1982. Much of this came from refining the production system and changing the plant layout in so doing.

Block Ⓓ: The five-year V–2 campaign had many elements, but it stressed a great deal of what Americans would call value engineering coupled with industrial engineering, a very straightforward type of program. Toward the end of this period it began to use more and more of the concepts of stockless production. Tachikawa Spring is continuing with this, and intends to employ stockless production to guide the introduction of robots.

EXHIBIT 2–5 *(concluded)*

	71	72	73	74	75	76	77	78	79	80	81	82
(A) Indexes of improvement 1) Added value productivity* 2) Sales volume 3) Total wages 4) Labor relative share† 5) Number of active part numbers		1) 1.0 2) 1.0 3) 1.0 4) 1.0			5) 1.0		5) 0.3	1) 2.5 2) 3.1	3) 2.3 4) 0.93			1) 3.3 2) 4.2 3) 3.0 4) 0.94 5) 0.3
(B) Labor productivity index Inflation-adjusted sales Total employees		1.0									1.97	2.06
(C) Reduction of inventory increased flexibility (about 40% reduction in space needed)							6.0	4.5	11 min. 47 times per min.			7.8 min. 57 times per min.
(D) Tachikawa Spring production programs			"V-2" campaign: Value added productivity (73–78)						Refine Tachikawa Spring production system (79–82)	Lay out plant by customer models (80–82)	Introduce robots (81–82)	

Chart labels: Added-value productivity per employee; Number of active part numbers; Days-on-hand manufacturing inventory; Press set up times; Press set ups per month.

$$\text{* Added value productivity} = \frac{\text{Sales} - \text{Value of purchased materials and services}}{\text{Number of total employees}}$$

$$\dagger \text{ Labor relative share} = \frac{\text{Total wages}}{\text{Added value}}$$

EXHIBIT 2–6 Development of stockless production at Jidosha Kiki Company Ltd.

	Start of program 1976	*1981*	
Inventory (days on hand)			
Raw material	3.1	1.0	
Purchased parts	3.8	1.2	
Work-in-process	4.0	1.0	
Finished goods	8.6	3.7	
	19.5	6.9	
Productivity index	100	187	
Defect rates			
From suppliers	2.6%	.11%	
Internally (cumulative)	.34%	.01%	
Setup times			
Over 60 minutes	20%	0%	
30–60 minutes	19%	0%	
20–30 minutes	26%	3%	
10–20 minutes	20%	7%	
5–10 minutes	5%	12%	Single-digit setups
100 seconds–5 minutes	0%	16%	Single-digit setups
Under 100 seconds	0%	62%	One-touch setups

This data was taken from a briefing by Jidosha Kiki executives during a visit July 2, 1982. Jidosha Kiki, which supplies brakes and steering gear to the auto industry, learned its methods from Toyota, and their system is very similar. They began a very clear conversion to this kind of thinking with a "revolution" in 1976, and the results have been very good for them. Jidosha Kiki has a few large presses. Most of the rest of their equipment is small, light, flexible metalworking equipment.

The extremely low amount of raw material, purchased parts, and finished goods results from the truck delivery system of Japan, described later. Jidosha Kiki executives were unsure if these levels would be further reduced, but they were certain that additional in-plant improvement could further reduce work in process to 0.5 days on hand.

Most project leaders are enthusiastic, but naturally there are many other people in each company who must see some proof before fully committing themselves and their companies.

Many of the American projects include only one or two of the concepts of stockless production. Projects to reduce setup times are popular, and early results are promising. For example, one of the plants of the Brunswick Corporation reports that setup times which once took 16 hours are now done in 2 hours, and setups that took 2 hours now

EXHIBIT 2–7 Comparison of manufacturing inventories*

Company	*Days on hand*	*Annual§ turnover*
Toyota Motor Company† (1980)	4.0	62
Tachikawa Spring Company‡ (1982)	3.3	75
Jidosha Kiki‡ (1982)	3.2	78
Kawasaki Motorcycle—Japan (1981)	3.2	78
Kawasaki USA (active parts—1982)	5.0 est.	50 est.
Tokai Rika†,‡ (1982)	3.7	68
American competitors (1981)	10–41 est.	6–25 est.

* Manufacturing inventory is defined as raw material, parts, and work in process. It does not include finished goods. The companies using stockless production typically count the manufacturing inventory at the end of each month. Estimates of the American companies were made by materials managers of the companies at a meeting of the American Production and Inventory Control Repetitive Manufacturing Group in June 1981. The estimate of days on hand at Kawasaki USA was made by the inventory manager.

† Figures include inventory on consignment at small suppliers.

‡ These companies are among the larger suppliers of the Japanese auto industry.

§ Annual turnover and days on hand are related by

$$\text{Turnover} = \frac{(\text{250 Days per year})}{(\text{Days on hand})}$$

The Japanese companies typically state inventory in days on hand, the American ones by a turnover ratio. American reaction at first is that the amount of money required for inventory is minuscule. That is true, but recall that inventory levels are really an indicator of the degree to which the production process itself has become flexible and free of breakdowns and rework. The financial impact of that is enormous, but it is not explicitly identified as a separate item in the financial statements.

require only half an hour. Reducing setup times appears to be mostly a matter of concentrating on it.

The principles by which equipment operates and material moves are the same everywhere, and the general ideas of stockless production are valid everywhere. A recent survey by Professor Roger Schmenner of Duke University lends credence to the thought that the

EXHIBIT 2–8 Development of the supply network for Toyota (as shown by days-on-hand inventory)

	Tokai Rika			*Nippondenso*	*Jidosha Kiki*		
	1975	*1982*	*Percent change*	*1981*	*1976*	*1981*	*Percent change*
Raw materials and parts	—	2.5	—	0.9	6.9	2.2	−68%
Work in progress	8.0	.75	−90%	4.0	4.0	1.0	−75%
Finished goods	9.1	1.80	−80%	3.0	8.6	3.7	−57%
	—	5.05	—	6.9	19.5	6.9	−65%

Each of the three companies above are major suppliers to Toyota Motor Company, three of the 225 suppliers of production parts, and members of the Toyota family of suppliers. All three began programs of stockless production in the mid-1970s, when Toyota began instructing their suppliers in the system. These companies are regarded as among the better of Toyota's suppliers, but there are many others equally as good.

In each company there was a healthy decrease in overall inventory, but notice that the greatest decrease was in work-in-process inventory, which indicates that a great deal of attention was paid to improving operations and streamlining material flow inside the plants.

The decrease in raw material and finished goods inventories was also remarkable. The numbers are amazingly low by American standards. That is partly due to the short distances between plants, which occurs not only because Japan is a small country but also because decisions on plant location are often made with short transit distances in mind.

However, the companies themselves did not make major changes in the location of plants during the periods shown. The decrease in raw material and finished goods was made because the improvements inside plants resulted in much more stable, regular schedules. That allowed the suppliers to connect their own operations more closely to Toyota (and to other customers that each of them supplies). It also allowed the interplant transport of parts to be made on a more regular schedule, thus improving the efficiency of transport itself; but the most important point is that the improvement of production operations everywhere allows the industry to operate with less safety stock throughout the entire industry.

general direction is right for the United States. He surveyed manufacturers in North Carolina to determine what practices seemed to be correlated with improvement in productivity. In a paper he presented at the Academy of Management Annual Meeting in New York in August 1982, he said

> For the most part the independent variables which are highly correlated with productivity define classic ways of simplifying and smoothing out the production process. For example, it was found that many engineering change orders are disruptive to productivity as are frequent design changes and product mix changes. Smoothing the flow of material is important. Those processes which had improved material handling, reduced scrap, lowered rework, improved quality, lessened expe-

EXHIBIT 2–9 Booze-Allen-Hamilton comparison survey of inventory turnover ratios

Table A: Japanese versus American inventory turnover ratios (1978)

	Japan	*United States*
Job shops	3.1	3.2
Repetitive manufacturing	7.6	3.8
Process industries	5.5	5.1

In 1980, Randy Meyer and Robert Fox of Booze-Allen-Hamilton surveyed 1,500 manufacturing companies in Japan, Europe, and the United States. The companies were placed in one of 15 different industries according to SIC code or the equivalent, and for the table above, a judgment was made as to whether each company was more job shop, repetitive, or process in most of its operations. The turnover ratios were taken from financial statements sometimes corroborated by personal contact with the companies. The companies were randomly sampled without regard to their status with respect to stockless production. The complete report from which the data comes is confidential to Booze-Allen-Hamilton, but this data from it was provided through the courtesy of Meyer and Fox.

Of course, the most distinctive outcome is that the average turnover ratio of Japanese repetitive manufacturers was exactly double that of their American counterparts. The difference in other types of industries is not so startling. The data is now four years old. In 1978, stockless production was beginning to take hold in the leading Japanese repetitive manufacturers.

Table B: Japanese versus American inventory turnover ratios in four selected industries

	Japan		*United States*	
	1970	*1978*	*1970*	*1978*
Auto and truck	9.2	11.9	4.3	5.6
Auto parts	6.3	11.3	3.2	3.9
Electronics	4.6	6.0	3.6	3.2
Construction equipment	2.8	3.0	3.7	3.2

This data was developed from the same Booze-Allen-Hamilton survey as that in Table A. During this period in Japan, the leading auto plants were struggling to keep abreast of Toyota, but performance at several of the smaller companies might not have been nearly so good. The degree of improvement in auto parts reflects the introduction of stockless production among Toyota suppliers and the beginning of action among suppliers to other auto companies. The Japanese electronic companies were just beginning to think about stockless production, and most of their improvement programs concentrated on higher yields through quality control and on increased automation through more conventional manufacturing engineering. The production of most construction equipment is not repetitive, and the Japanese industry was just beginning to consider the improvement of shop floor organization and how they might employ Material Requirements Planning (MRP) to assist them.

EXHIBIT 2–10 Implementing stockless production at a small alloyed aluminum wire plant of Sumitomo Electric Company

This plant is a small subdivision of Sumitomo Electric Company—less than 30 employees. It processes and coats 200–300 tons per month of aluminum wire by customer order. Customers want delivery with short lead times, usually two to three weeks, and of the 200 or so orders per month, the majority are small (less than 1,000 pounds.) The workload is difficult to forecast. The plant cannot operate by a smooth flow, so it plans and controls work by a job shop system.

Prior to stockless production, the plant worked intensively on improving the reliability of the process itself. Formerly, the scheduling consisted of a monthly plan, which was then rescheduled by an expediter as the daily situation required.

The monthly planning and rescheduling were abolished. Now each order is scheduled as it is received. Starting dates for each process required by an order are backscheduled from the delivery date. Work load is considered when promising the delivery date, and the backscheduling is based on standard flow times for each process. The large lots are scheduled in segments so as to provide more flexibility for all orders in scheduling and also to level the plant load.

A computer generates a machine load projection daily for the next four weeks. Then printed cards and a schedule tape for each order are used to manually show the schedule on a schedule board in the middle of the plant where all workers can see it.

The key changes that make the system work: The excess work in process has been removed from the floor, so work can be easily tracked and monitored. The schedule board is visible to everyone. Workers can refer to it to determine for themselves when they need to work overtime. They also do their own posting to the board to show how actual operations compare with schedule. The small size of the plant and the physical improvements have made problems visible as they develop.

This situation is quite different from the approach to stockless production for repetitive manufacturing that will be described in detail. It is included here to illustrate that the basic principles are broadly applicable to a variety of industry.

This brief history is taken from "Stockless Production," an unpublished report from Sumitomo Electric Company, by Dr. Ryuji Fukuda, who was manager of industrial engineering when he wrote it in 1980. (Some time after the author began to refer to the system as stockless production, the report by Dr. Fukuda was discovered. It was found that a number of Japanese call it stockless production, although that is not one of the most popular names for it.)

diting, lowered production cycle times, reduced shortages, adopted MRP systems, or had significant numbers of their workers cross-trained appear to have a higher level of labor productivity than plants which have not enjoyed this kind of change. Those plants which measure productivity on a units-per-worker-hour basis also tend to have higher labor productivity than others, although significantly, plants that have either individual or group incentives do not appear to have higher labor productivity than plants that pay by the hour.[1]

Schmenner uses the phrase *production cycle times* to mean what in this book is called throughput times or lead times. Every factor he

[1] Roger W. Schmenner, "Strategy Elements of Factory Productivity" (Paper presented at the Academy of Management Annual Meeting, New York City, August 16, 1982).

EXHIBIT 2–10 *(concluded)*

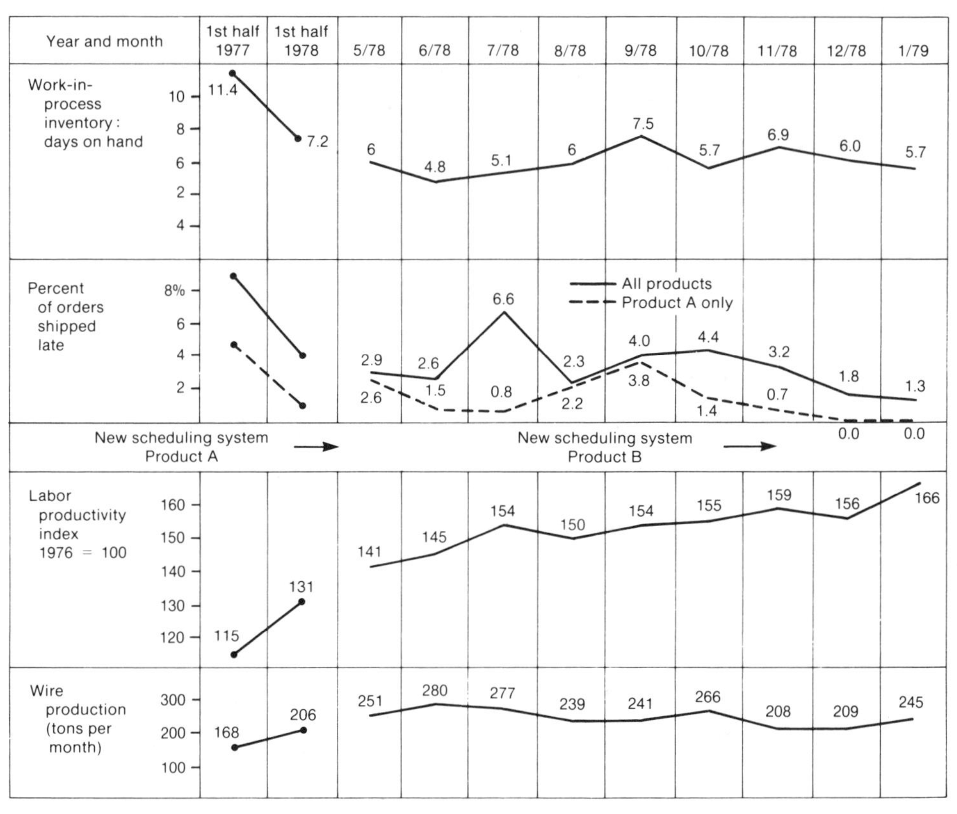

Source: Dr. Ryuji Fukuda in "Stockless Production."

cites as being correlated with productivity in North Carolina is consistent with a concept of stockless production which has already been presented or which will be presented, although stockless production would not advocate using the typical job shop form of MRP for production which can be made repetitive. What Schmenner found was that it is the basics which count, and he did not have stockless production in mind when making the survey.

Much of stockless production is a reaffirmation of basics, some of which may be half forgotten. The advice from those who have done it before is to first get the basics firmly in mind, then learn the details by doing it.

EXHIBIT 2–11 Results of stockless production projects in place in American plants, October 1982

Company and location	*Nature of project*	*Inventory reduction*	*Results*
General Motors—Pontiac	Level schedule of intake manifold. Some physical improvement.	About 70 percent work-in-process reduction in nine months.	Have obtained perfect quality rating by GM corporate "several" months in a row.
General Motors—Oldsmobile	Move bumpers to assembly by card.	About 65 percent lower bumper parts bank.	Forcing improvements in bumper production, but not yet quantified.
Westinghouse—Bloomington	Use card system to pull parts for assembly of throw-out switch.	About 60 percent in one year ($30,000 reduction).	Approximately 25 percent increase in labor productivity. Quality improvement. Line stops reduced to two to three per month from a previous, "large" number.
Hewlett-Packard—Vancouver	Issue quantities daily to two assembly areas. Parts drawn and replaced by circulating marker system.	About 30 percent in three months.	About 35 percent reduction in assembly hours. Less rework in final test, but no quantification of the estimate.
Hewlett-Packard (Engineering Systems Div.)—Ft. Collins	Subassembly and final assembly parts drawn by signals from assemblies.	Just started. Expect to need no more than 50 percent of previous inventory.	Quickly accepted by workers. Plan to extend system to supplying area and to some vendors.

Americans are just beginning to experiment with stockless production. The number of such projects in the planning stage is very large, and some are known to be in operation, though the author has not seen them. These are only the projects personally seen by the author as of October 1982.

3

Identifying problems through the production control system

Most of the manufacturers in the world create production schedules on the assumption that they can be executed. If work is completed as scheduled, it is sent from a work center to the next location at which parts are scheduled for use. This is a "push system": Make the parts and *send* them to where they are next needed, or to inventory, thus pushing material through production according to schedule.

In this system, the function of production control is to keep production on schedule. Most manufacturers have deviations between what was scheduled and what actually occurs, so production control consists of recognizing these deviations and taking action. Three basic kinds of action take place in response to deviations from schedule:

1. Adjust activity to get back on schedule.
2. Expedite.
3. Reschedule.

Of the three actions, the last two are not considered desirable. Their use is an indicator that the company was either not able to develop a valid schedule or to execute it if it was valid. We usually *assume* that we should be able to schedule and control a plant as it exists—somehow.

Turn this assumption on its head. What are the real reasons we miss schedules? Need all the inventory assumed by the schedule? Require slack time in the schedule? The reasons involve all the kinds of problems discussed in Chapter 1, but it is difficult to sort out what to do. Begin to work on the more obvious problems and use the production control system itself to reduce inventory levels and identify others we should work on.

The general idea can be applied to any form of manufacturing, but the system for it has been most completely developed when applied to repetitive manufacturing. This involves stabilizing and smoothing the final assembly schedule, then pulling material into final assembly.

What an expediter in a job shop frequently does is pull material into final assembly. He goes to a shop floor crowded with inventory and decreases the interruptions in the production rate of the critically needed material. The priority parts move while the others sit, and the expediter tries to arrange all operations so that the critical parts flow into final assembly as quickly as possible. He temporarily converts the job shop into a repetitive manufacturer, but only for the priority parts, leaving a mess of the schedule in his wake.

Suppose a manufacturer wanted to expedite everything, and do so without making havoc of a schedule. It cannot be done economically by having a great deal of extra capacity; so the plant, the schedule, and the control system must be developed for it. For production to flow smoothly, we need to eliminate scrap and machine breakdowns. In order for all material to go through production as quickly and smoothly as possible, we want to avoid having inventory accumulate at bottleneck points in production. In short, we want to do those things envisioned by stockless production.

In order to make production go smoothly, we would like to convert as much job shop production as possible to repetitive manufacturing. This may require a great deal of physical revision to the equipment, including reduction of setup times, but from the viewpoint of production control, it may not be as impossible as first thought. Many manufacturers (and entire industries) operate by scheduling the fabrication of parts into an inventory, from which parts are drawn for final assembly. Scheduling and controlling the fabrication are done by a push process except when expediting. However, for final assembly, drawing parts is more like a pull process. Material is drawn as needed. The idea is to begin to draw parts to final assembly directly from fabrication as much as possible. To make this work well, the planning system for it works basically as follows:

Plan a level final assembly schedule in advance. Level in this case means to make a little bit of everything every day—a little bit of everything every hour if possible—assembly lot sizes of one.

This helps to balance all the operations feeding final assembly. Individual parts are not needed in large surges.

Develop a master production schedule based on the final assembly schedule. The master production schedule is a summary in daily buckets of the final assembly schedule. There are two cases of this:

Make to order or assemble to order.

Make to stock.

Explode the master production schedule using bills of material to derive level schedules for fabrication and subassembly operations which feed final assembly. If the final assembly schedule requires a level, even usage of parts throughout each day, then the requirements for parts will be level throughout each day.

On a daily or hourly basis, adjust the actual final assembly schedule to reflect the deviations from plan caused by:

Changes in customer orders from what was planned.

Production problems.

Provide a pull system of production control to bring parts through the production operations to match the demands made by the actual final assembly schedule. This system should:

Get the right part to the right place at the right time.

Prevent the ballooning of throughput times by controlling the amount of work-in-process inventory for each part at each operation.

Prevent the need to put inventory in stockrooms.

WHAT IS A PULL SYSTEM?

The term *pull system* is used in many places in this book. It means that material is drawn or sent for by the users of the material as needed. The idea was phrased very succinctly with intentional lack of grammar by a General Motors foreman who said, "You don't never make nothin' and *send* it no place. Somebody has to come get it."

This has a great deal of merit as a way to see what parts are really needed and which ones are not. If completed production is stacked at the point where it was made and no one comes for it, the workers can readily see that production of that part should stop. If it is sent to a stockroom or somewhere out of sight, it takes a little longer (and perhaps reference to an inventory report after a delay time) to make the same decision. On the other hand, if someone wants parts that have not been made, that message is also clearly and immediately understood. The result is that people want to have on hand whatever will be wanted, but not too much, especially if space for parts is limited.

Think of a pull system as operating on stock in a pipeline, which connects the work center where the stock is made with the work center where it is used. Then for stockless production to work properly, the pull system should have two characteristics:

1. It must synchronize the movement of material to the rate at which it is withdrawn for use at the output end of the pipeline, thus allowing for market deviations from the planned schedule.
2. It must constrain within some limit the total amount of inventory in the pipeline.

If the inventory in the pipeline is kept within a limit, the result is a fixed-volume pull system (Exhibit 3–1). This is very important in con-

EXHIBIT 3–1 Concept of fixed-volume pull system

If the feeding process has nowhere to put inventory except in the pipeline, it can only produce and fill the pipeline when an empty space appears.

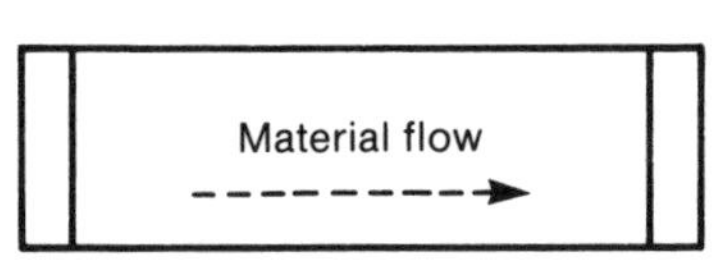

Removal of material for use causes an "empty space" in the pipeline. Material moves down the pipeline to leave an empty space at the feeding end.

trolling the level of stock at various points and for making controlled reductions of pipeline stock.

As the amount of pipeline stock is reduced and the frequency of removal from the output end increases, the effect ceases to be one of controlling an inventory system at all. The removal of stock for use becomes a *signal* to action on the other end in order to keep the users supplied. This point is very important for advanced development, but difficult to see at first.

The Toyota card control system is a fixed-volume pull system. There may be other ways to do it, and certainly there are other pull systems. The broadcast system of the auto industry is a pull system, but not always fixed volume.[1] However, for stockless production, many companies have now copied the Toyota system, and some have added their own variants. It illustrates such a system in detail.

THE TOYOTA CARD CONTROL SYSTEM

Because Toyota calls their entire system Kanban, people sometimes imply a power to the cards that is not present. It takes total

[1] A discussion of the broadcast system is in Chapter 8.

development of production for the card system, or any system of signals similar to it, to work. In Japan, signs, posters, and tickets are all called kanban, but within Toyota the term refers to a move card or to a production card.

Before the card system can be put into place, substantial physical revision of the equipment and plant layout are necessary. Routings must be defined and fixed so that each part has a clear path through production, and so that in each plant there is only one source location for each part.

Each work center must be defined and organized so that inventory is held only at the work centers and not in stockrooms. Each work center has an inbound stockpoint and an outbound stockpoint. Assembly lines have one or more inbound stockpoints that serve as staging areas, where material is organized for positioning in exactly the right location for easy access during assembly work. In effect, the entire plant becomes organized something like a stockroom. That is necessary if all active inventory is to be kept on the plant floor without confusion. Exhibits 3–2, 3–3, and 3–4 illustrate the development of the plant floor to use the card system. Note that suppliers are considered to be work centers which happen to be shared with other companies, and the card control system extends to them just as if they were inside the plant. Two kinds of cards are used:

Move card. This card (Exhibit 3–5) authorizes the movement of one part number between a single pair of work centers. The card circulates between the outbound stockpoint of the supplying work center (where the part is made) and the inbound stockpoint of the using work center. The card is always attached to a standard container of parts when it is moved to the using work center. The information on a move card includes:

Part number.
Container capacity.
Card number. (Example: No. 4 of five cards issued.)
The supplying work center number.
  That work center's outbound stockpoint number.
The using work center.
  That work center's inbound stockpoint number.

Production card. This card (Exhibit 3–6) authorizes the production of a standard container of parts to replace one just taken from an outbound stockpoint. These cards are used only at the supplying work center and its outbound stockpoint. Information on a production card:

Part number to be produced.
Container capacity.

Supplying work center number.
  That work center's outbound stockpoint number.
Card number. (Example: No. 2 of three cards issued.)
Materials required:
  A short bill of material for parts needed.
  Outbound stockpoint locations from which to get these parts.
  Other information, such as tools needed.

EXHIBIT 3–2 Layout for stockless production

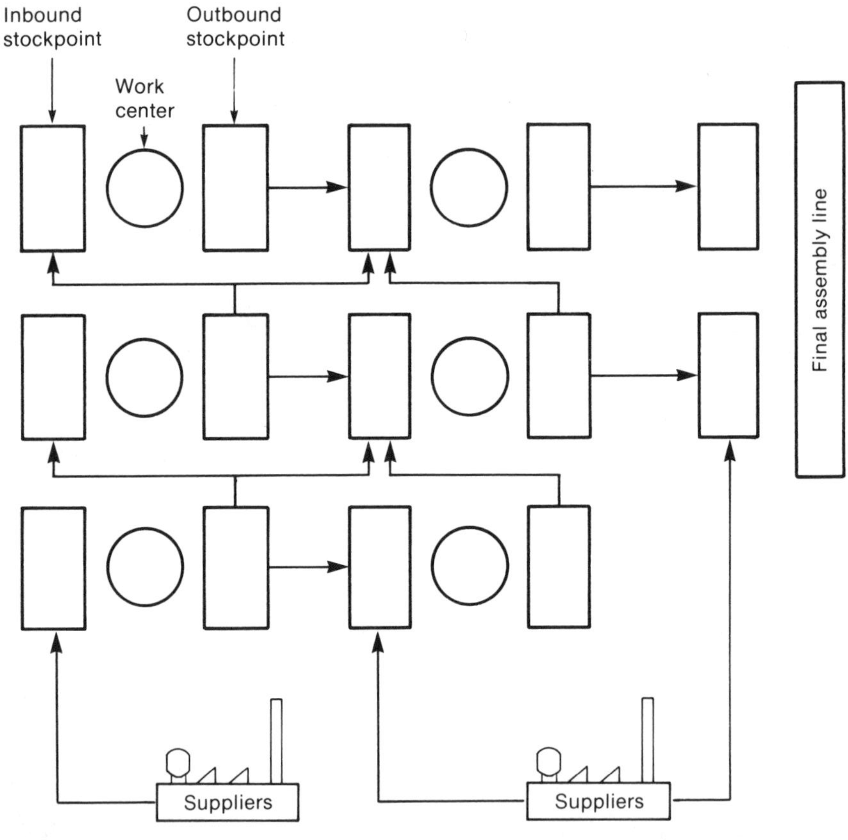

Stockless production requires reorganizing the plant to eliminate stockrooms for production inventory. All production inventory is held on the production floor at work centers organized as shown here, each with an inbound and an outbound stockpoint. The arrows show material flows going from outbound to inbound stockpoints. Although there are six work centers shown, the routings are all fixed-path, straight-through material flows.

EXHIBIT 3–3 Flow paths of cards (signals)

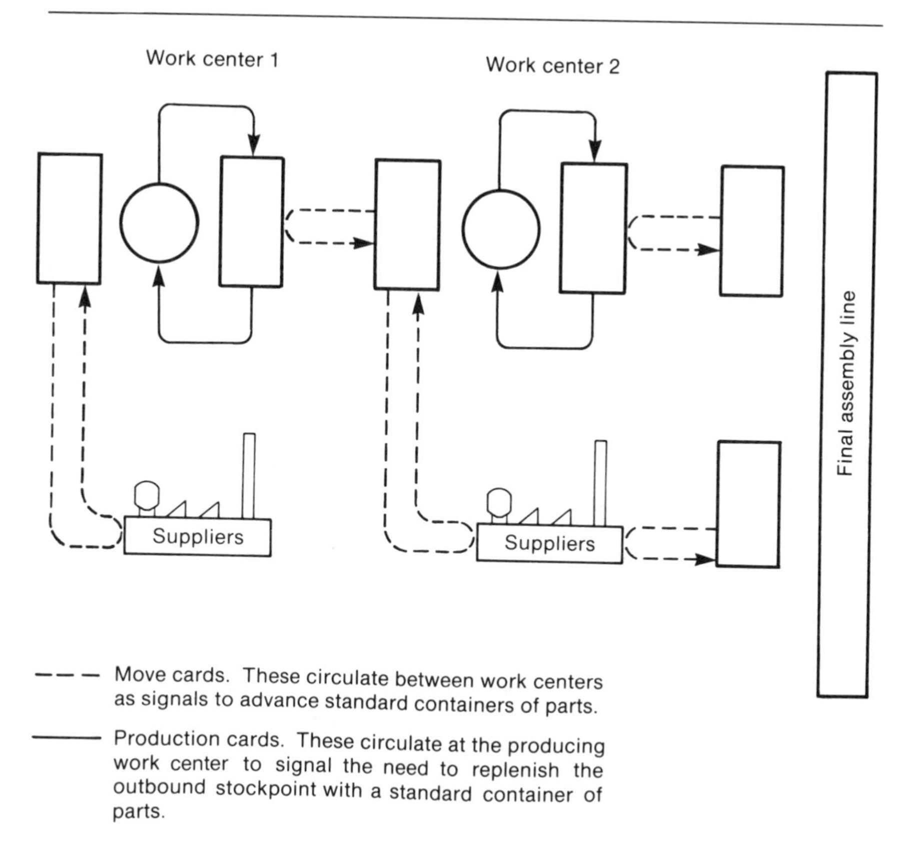

With the system simplified to two suppliers, two work centers, and one final assembly line, we can see paths of the cards (or other signals) that keep the pull system going.

When a container of parts is selected for use from the inbound stockpoint, the move card is detached and placed in a collection box. It will then be picked up and taken back to the supplying work center as authorization to get another container of parts. Move cards, therefore, circulate between only two work centers. Each designates only one part number.

When a move card is brought to an outbound stockpoint to get parts, the production card is removed from the standard container selected. The move card is attached to the standard container, and it is transported back to the inbound stockpoint of the using work center ready for use.

EXHIBIT 3–4 Kanban cards used as pull signals

Using only two work centers, the paths of the cards can be shown in more detail:

■ Move card — – – – Move card path

□ Production card — ——— Production card path

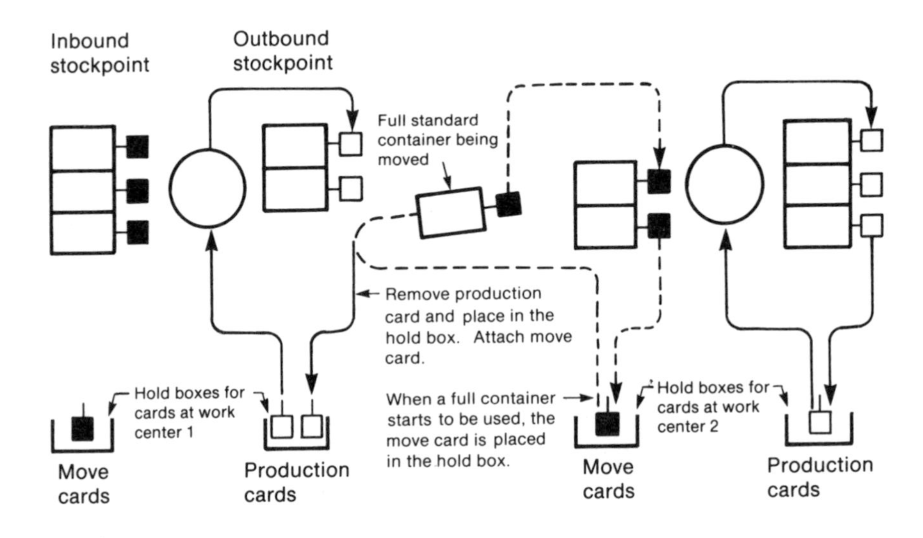

EXHIBIT 3–5 Typical move card

FROM SUPPLYING WORK CENTER #52 PAINT	PART NO. A575 GAS TANK MOUNT	TO USING WORK CENTER #2 ASSEMBLY
OUTBOUND STOCKPOINT NO. 52-6	CONTAINER: TYPE 2 (RED) NO. IN EACH CONTAINER: 20 CARD NO. 3 NO. CARDS ISSUED: 5	INBOUND STOCKPOINT NO. 2-1

EXHIBIT 3–6 Typical production card

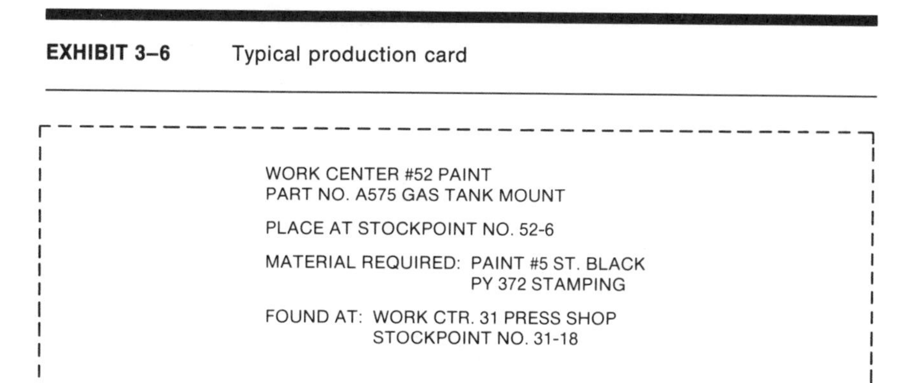
WORK CENTER #52 PAINT
PART NO. A575 GAS TANK MOUNT

PLACE AT STOCKPOINT NO. 52-6

MATERIAL REQUIRED: PAINT #5 ST. BLACK
PY 372 STAMPING

FOUND AT: WORK CTR. 31 PRESS SHOP
STOCKPOINT NO. 31-18

EXHIBIT 3–7 Move card used by Hino Motors Ltd. to pick up parts from a supplier

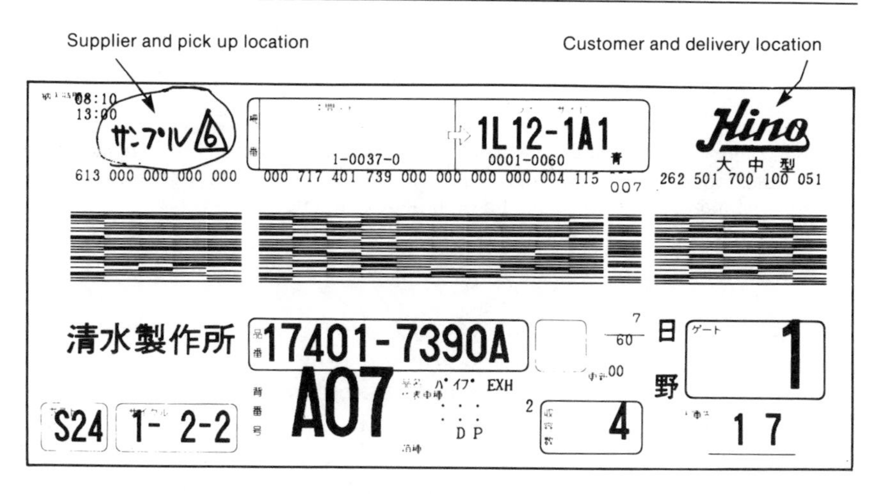

Note the bar code. Only move cards sent to suppliers are imprinted with the bar. Upon receipt, the bar code is read, and a cumulative total is kept on-line by computer. Hino uses it only for development of receipts to pay the supplier's invoice, an accounting function not related to control of the part for production. However, the cumulative counts can be used by production control to keep a cumulative count of parts received.

The times posted at the upper left refer to delivery times, twice daily for this part. Sometimes the times posted refer, more precisely, to the times the move cards must be picked up to give to the truck driver bringing a new delivery of parts.

EXHIBIT 3–8 Stockpoint organization diagram

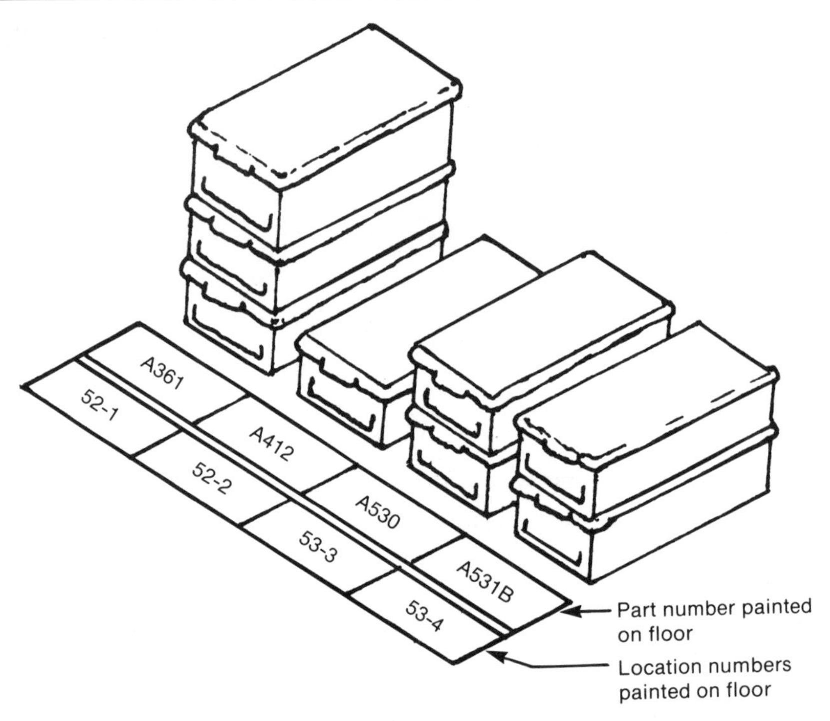

An outbound stockpoint supplying several parts might be organized by marking the spot on the floor or table on which each part number may be found. For many parts, the standard containers can be stacked on the floor at the designated spot.

Numbers may be painted on the floor. More likely they are painted on signs that rise above the floor.

The production card which was removed is placed in a collection box at the supplying work center. From time to time workers there collect these cards. Each production card collected is authorization to make another standard container full of parts to replace one that was just taken.

The work rules using the kanban cards are simple, but strict:

1. Either a move card or a production card must always be attached to containers holding parts.[2]

[2] There is an occasional exception to this when production cards are reduced in number at an outbound stockpoint, either at the time of a schedule change or just to reduce pipeline stock. At that time, workers should *first* take any standard containers without cards before taking those with production cards. Otherwise, the work center will start to replace stock before the "excess" inventory is cleaned out, and the discipline of a fixed pull system will be broken.

2. Using work centers must always come get the parts from supplying work centers, or signal the supplying work centers for parts using the move cards. Never transport a container of parts without a move card.
3. Standard containers are always used. Never use nonstandard containers or fill the standard containers with a nonstandard number of parts.
4. Only produce a standard container of parts when an unattached production card authorizes it. Always attach the production card to the full container when it is placed in the outbound stockpoint.

If these rules using the cards are followed, the system provides a very simple pull system by which all material is synchronized in moving forward from raw material to final assembly.

The pull system always starts from the final assembly line or final assembly area. The production control department gives the final assembly schedule to the final assembly area. It may be the same as the final assembly schedule originally planned, or it may deviate slightly from that schedule. For example, the auto industry assembles units only to order. They may receive order cancellations and need to make detailed adjustments in the final assembly sequence run. As long as these deviations do not become more than, say, a 10 percent change from the original plan on a given day, there is little trouble pulling the right parts to build the sequence of units assembled.

All workers in the system have the obligation to provide parts to those who need them. Everyone must be able to come get what they need when they need it. This is most important because it overcomes the psychology that one must hoard parts to protect against the vagaries of the unreliable people who provide them. Therefore, the running of a schedule that is very near to that originally planned is important, and all workers must produce what is required to be produced. Only then does the system hold together.

With the proper disciplines in place and with the plant revisions made to accomplish a direct flow without a stockroom, production control proceeds as follows:

1. To meet their schedule, final assembly workers withdraw parts as they need them from the work centers that supply them.
2. Since many parts come from suppliers to final assembly, as a practical matter many of the cards are collected and given to the truck drivers delivering from suppliers. These cards returning to suppliers are the request for the amount to be delivered with the next load.
3. The work centers supplying the final assembly line make just enough parts to replace those that have been taken.

4. Continuing in this way, each work center withdraws parts as needed from the work centers supplying them. This continues in a chain all the way back to the suppliers.
5. Much of the move card transport is actually performed by the material handlers. The workers use the cards as a signaling system.

USE OF THE CONTROL SYSTEM TO STIMULATE PROCESS IMPROVEMENT

It takes a great deal of plant development just to get to a point where a pull system will work in a very basic way. Much of this is guided by making reductions in inventory and studying in detail what has to be done, but the reductions are not made in a minutely controlled way. The pull system using cards can do this in a more precise fashion.

The pull system should be capable of controlling the inventory level of any part number at any point of use on the plant floor. Each part should be made at only one work center, but it may be used at several. Therefore the number of production cards in use is set to allow the producing work center to operate so as to barely be able to cover the needs of the using work centers when they come to get parts.

As a practical matter, most of the inventory should be held in the outbound stockpoints, but this may not always be the case: The amount of inventory held at the inbound stockpoints and in transit reflects the state of development of material transport and also the need which the using work center may have for parts before it can be resupplied. The number of move cards issued for each part number between each pair of work centers covers this.

Since one card goes only on one standard container, the total number of cards issued for each part number is proportional to the amount of work in process (or pipeline stock) allowed for each part number. This is further subdivided according to how many production cards are issued to the producing work center and how many move cards are issued for each work center pair. The inventory can be adjusted at any point in production by the following method:

1. Issue the cards at the start of a frozen schedule period (a 10-day to 30-day block, depending on the company).
2. Withdraw any cards that allow any inventory in excess of what is needed to buffer the imbalances between the supplying and using work centers for a specific part. This often is done by department managers.

3. Withdraw one or two more cards from the system and analyze the cause of the problems that occur. If you wish to study the problems of the producing work center, withdraw production cards. If you wish to study problems of material handling methods or schedules or of a using work center, withdraw move cards. What cards are pulled is a function of specifically what we wish to study and improve.
4. Of course, when cards are initially withdrawn from a pipeline in which stock barely allows coverage of the operations, everyone has to do some extra work, usually overtime, but this is not the real problem.
5. The "real" problems are sometimes anticipated, and sometimes they are revealed by the inspiration provided by the extra pressure from reducing the inventory. The expectation is that a specific improvement or two will emerge from the muddle. A new idea to further reduce setup times? Change machine or preventive maintenance procedure to reduce downtime? Better balance between production rate and use rate?
6. Get everyone involved in thinking of a solution. Deliberately withdrawing cards focuses the attention of both workers and supervisors on the problems.
7. Make the changes that are easiest or least expensive in order to operate smoothly again at the new lower level of inventory. Changes may be in work methods, equipment, or scheduling.
8. When the flow of parts is going smoothly again at the new lower level of inventory, withdraw one or two more cards and go through the process again.
9. Go through the process again. Keep going through this cycle until no cards are required at all, so there is no need for inventory between processes. Of course, one may also be temporarily stymied because no economical way comes to mind to further reduce inventory between two operations.

This is the true use of the system. Make problems visible so that many people can see them. Develop organized ways to attack the problems as they are revealed. As soon as production is running smoothly, tighten the flow so that people never become complacent, but are always working on as many improvement problems as they have the capacity to handle. Attaining a pull system of control is not an end in itself. It is only a method for promoting a relentless pursuit of stockless production—eventually full automation. Management should never be satisfied with its current condition, no matter how advanced. The existence of a pull system does not reveal problems by itself. It is the reduction of inventory which does it.

It is the obligation of line management to see that inventory is kept low enough at all work centers that everyone has to concentrate on the process and how to improve it. No one has the luxury of relaxing while others struggle. Otherwise there is resentment and dissension among the floor work force.

VARIATIONS IN THE CARD SYSTEM

There are many variations of this system. If the same part is made by the same people every day, the information on the cards soon becomes unnecessary. Pull signals are therefore given by colored washers, colored chips, plastic markers, electronic signals, colored golf balls (transmitted through a pneumatic tube), and the containers themselves. The containers may be only colored, or they may be tagged with much more information. In modifying the system, it is important that it should easily provide the function of keeping a limit on the pipeline stock as well as providing demand signals (not always done when practice becomes sloppy.)

It is very common for the production cards not to be used, although this can lead to difficulty in keeping an upper limit on the amount of stock to be presented for each part at a supplying work center's outbound stockpoint. The workers follow a simple rule such as, "Stack no more than four containers of part Z876 at its stockpoint." If someone comes and takes two from a stack of four, they know to make two more.

When using this one-card system, the Toyota family companies often use a warning system with a triangular-shaped marker, as shown in Exhibit 3–9. These markers themselves have the part number painted on, and they slip between the boxes stacked in the outbound stockpoint. When the box on top of the signal marker is removed, the marker is placed on the tree, which is in a very visible location so workers are constantly aware of which parts are low. It tells them that if they are not already running that part or setting up for it, they should be. (A marker of this type, or a set of production cards placed in a visible place, is commonly referred to as *parts hanging*.)

One of the variations at Tachikawa Spring is in the press department, which makes large frame parts for vehicle seats, as illustrated in Exhibit 3–10. These parts go in large wheeled containers, not in small containers that can be hand-carried. Lanes are painted on the floor for the standard wheeled containers, one lane for each part number. As the containers are filled, they are wheeled into these lanes. Material handlers come get the containers for the work centers that need them according to a move card system, and the lane begins to empty.

Near the end of each lane is a warning line that shows about two hours of inventory are left for most parts—the signal that a press should be either running that part or setting up for it. They do not fill

EXHIBIT 3–9 Diagram of outbound stockpoint with warning signal marker

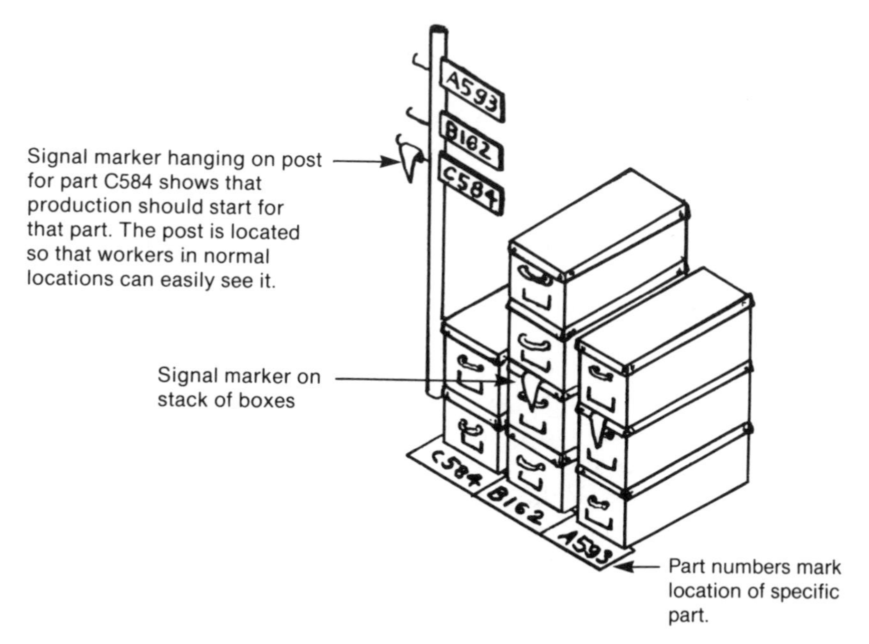

Note: These are more than reorder point systems. The intent is to progress beyond the use of these signals, not use them indefinitely.

EXHIBIT 3–10 Diagram of stockpoint lanes at Tachikawa Spring Company press shop

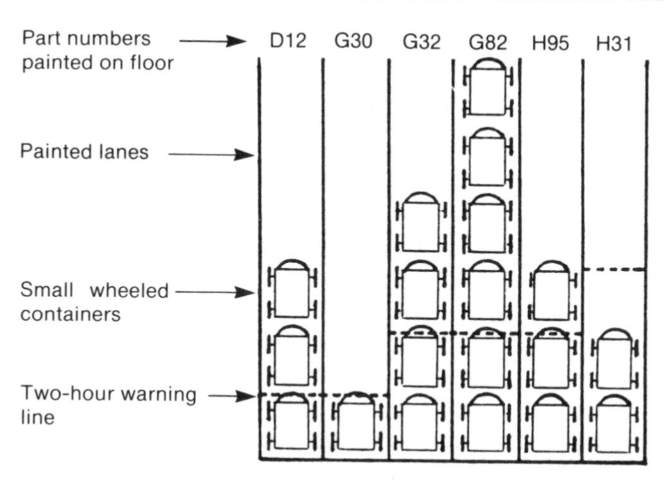

From the status shown in the diagram above, the lane has just been filled with part G82; they may be working on part H31 and getting ready for part G30, but that depends on the pull signals.

up enough containers to completely fill the lane for each part number. That depends on part use, how many go into a container, and how often they can set up for that part. (They appeared to be running about a half day's supply of most parts on one setup.)

Note that neither of these two examples of variations was referred to as a reorder point system, or as a two-bin system. Certainly they look like it, and there are many, many variations of coding methods similar to this which have been used in the United States and abroad. In fact, it is very difficult to keep the supervisor and workers from regarding these as just reorder point methods because the determination of the reorder point is done in the conventional manner. So what is the difference?

The difference is that management should always press to have the inventory reduced, and the response of each work center to the work centers it services should be so prompt that each work center using this variation of the system responds to signals of what is wanted by the users of the parts. If the workers only respond by rote to a reorder point, they may be making Part A because a withdrawal has triggered a signal marker, but the demand by the users has shifted to Part B. This requires a degree of alertness not present if people regard the system as "just a reorder point system." This is a difficult task even for companies well advanced in stockless production. What is desired is to continue improvement until a pull system which does not use the warning signals can be adopted, if possible.

In fact, the existence of the signal markers is evidence that much potential may remain for improving the methods of production. The next objective is to advance to using the card system or its equivalent without signal markers. Beyond that the objective is to not even use those but to remove inventory until cards are no longer necessary. Once the material flow becomes that smooth, direct transfer of material between work centers becomes feasible. The company is ready to begin working on the physical link problems of full automation.

Sometimes the cards are used to record production counts. The move cards are most often used in this way to count the receipts of material from suppliers. At work centers, if production becomes "as regular as clockwork," a mechanical counter can be used to record production. However, the more regular the production becomes and the lower the inventory becomes, the less important it becomes to record all those counts at many different points. Decisions are no longer made on the basis of requirements minus inventory.

A card system is not always a completely adequate pull system. The cards assume that all the parts represented by a card are identical, of course, but that is not always the case. For example, Toyota pulls engines into final assembly by a schedule that matches each engine with the correct vehicle. Since there are a great many variations in engines, the cards are not adequate to indicate exactly which engine "dress" variation must be sequenced with the vehicles being assembled. This is common to all auto manufacturers, who solve this problem by what the American auto manufacturers call a *broadcast system*. In Toyota's case, the cards for engines serve to keep the engine pipeline at a fixed volume.

A large company making a complex product must expect to have a lot of variations in the pull system. This is certainly true in the case of Toyota. They list 10 different kinds of "cards" in use in the system for various purposes. These are briefly described in Exhibit 3–11.

EXHIBIT 3–11 Variation in types of cards, markers, or signals used

A full-scope system may use a great variety of material pull signals. Here all of these are called cards. They are usually designed to look different from each other and thus are easy visual signals of what is happening.

1. Move card used internally between fabrication centers.
2. Move card used inside plants to bring material to final assembly.
3. Move card used to bring material from suppliers.
4. Move cards used by customers (their cards) to convey completed production.
5. Move card used for parts sent to stock. Usually used to accumulate spare parts at a stockpoint from which they are sent to a parts distribution system.
6. Signal card used to warn of the need to prepare for production of a specific part. Used in various ways: at outbound stockpoints or in a tool room, for example.
7. Production card. Not always used, but may be necessary if there is a complex flow of material within a producing work center, or if setups are difficult.
8. Move card used to draw standard items. Generally used for items such as fasteners.
9. Temporary cards. These are specially marked if cards need to be added to the system to cover temporary problems.
10. Special cards. Used for items such as
    Trial parts for new products.
    Engineering changes.
    Getting parts for test, engineering, samples, and so forth.

Note: Cards, markers, electronic signals, or whatever provides a clear signal may be used in a pull system.

As an example of the cards, consider No. 5—"Move card used for parts sent to stock." This one is often used for spare parts. Those spare parts made in sufficient volume to include in a regular production schedule are inserted in a uniform pattern, so many each day during each fixed schedule period. The parts are drawn from the work centers that produce them by material handlers who accumulate them in a shipping area or specially designated location, and from there ship them to the spare parts distribution system.

OTHER VISIBILITY SYSTEMS FOR PRODUCTION CONTROL

Visibility of floor status is attained in many other ways besides cards. Plants that feed many parts into one or two assembly areas have an electric "scoreboard" in the middle of the plant, visible to the maximum number of people. Identical scoreboards can be located at multiple points if desirable.

These boards, called *andons* in Japanese, generally give information to help coordinate the efforts of the linked work centers. The board posts the cycle time (time between units) of the units being assembled on each assembly line. Everyone can see which of the parts they are responsible for goes into each model and, considering that the parts are evenly dispersed among units being assembled, estimate whether they are producing faster or slower than the demand rate.

The boards may also have signal lights indicating the basic status of each work center. Green = running. Red = stop. Yellow = needs attention, or stopped for normal reason. The reason for showing this is

EXHIBIT 3–12 Tracking actual versus planned assembly

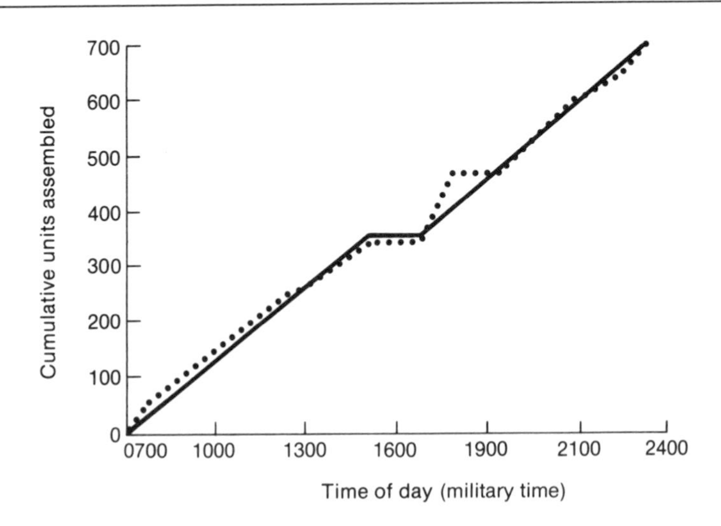

——— Planned assembly schedule. One hour between shifts for maintenance.

•••••• Actual assembly completions. This bulge would cause trouble. Drain the inventory from fabrication.

Note that the total deviation of actual from planned shows how much inventory is required to cover for the non-level performance.

so that workers can plan ahead on what the parts requirements will be. Most important, they should not continue running if the departments using the parts have stopped. If the final assembly areas stop and stay stopped, the fabrication areas also stop, sometimes without waiting to see that pull signals for material are no longer coming.

The scoreboard is very important for tracking actual assembly completions against the schedule. Where an electric board is not used, the same record can be posted on a blackboard or chart. Typically, these boards compare hour-by-hour cumulative completions for the day with what should have been completed by the schedule. Workers can tell where they stand for the day.

The objective of this is to have the assembly area actually perform by a level schedule. It does no good for a level schedule to be planned if the assembly workers cannot hold it. Therefore they should strive to meet the cycle rates for each model and to deviate as little as possible from schedule. Doing so insures that the parts requirements coming to final assembly are held as level as possible.

This is counter to the notion that workers should be allowed to pace themselves. Within some limits, they can pace themselves. They can stop the line if they are having trouble. They can speed ahead for a time, but if this is done for very long, they will outstrip the ability of fabrication areas to feed them parts. Therefore, the goal is linear production to schedule, indicative of level production in fact, not just by plan. The deviations of the final assembly area from the level schedule tend to be magnified as these cascade down through the feeding operations. Assembly workers need to understand that.

Establishing visibility of all forms of production problems is very important. Each operator has the responsibility to stop the process if defects are being produced, or if they are *about to be produced.* They are responsible for informing the preceding operation immediately if they are fed defective parts. (That comes naturally.) Instructions on assembly, tool care, quality checks are posted. Sometimes in plants where small parts are difficult to distinguish from one another, a sample part is attached to the standard container that holds it. Material handlers then know better what they are looking for, and operators can confirm exactly what it is they are to make.

The entire idea is one of instant communication. Records are not merely kept, but frequently posted for all to see what the problems are: defects and their causes, tool use, skills and qualifications of each worker. The purpose is to allow the people as a team to control the shop floor quickly and easily. If an individual tries to communicate well enough to control a complex production department, things are always limited by both his ability to acquire information and to communicate it. Therefore the principle is to communicate as much information as is possibly useful as soon as it is acquired, without delays for consultation or even for data processing unless absolutely necessary. Data and communication lags must also be covered by inventory.

As the plant becomes more automated it becomes possible for a single worker or a team of workers to service a large bank of equipment. They can be assisted by signal lights located directly on the machines as well as on scoreboards. This system allows workers to service a shop floor without much time spent figuring out what to do. Operators, maintenance people, and material handlers can all take their cues from the lights, scoreboards, and other signaling systems. The objective is to keep pressing toward that goal, but there is more to it than production control.

APPENDIX: CIRCULATING CONTAINERS AT MASTER LOCK CORPORATION IN MILWAUKEE

The use of cards or containers for signals in a pull system of material flow control is foreign to most American plants. However, Master Lock Corporation in Milwaukee has had a system of circulating containers for 35 years.

Master Lock makes padlocks and is best known as the company whose lock cannot be opened by a high-velocity bullet. They have about 1,250 hourly workers, organized by the United Auto Workers. The circulating container system coordinates three fabrication and subassembly areas with final assembly as shown below.

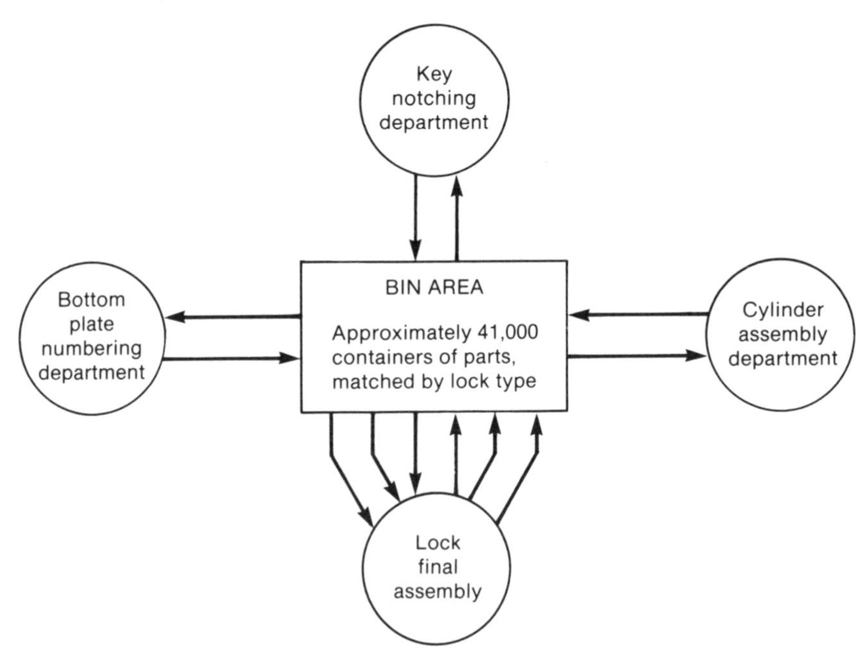

Locks are assembled in lots of 20. The matching sets of cylinders, keys and bottom plates needed for each lot are picked from the bins. The containers in the bins are color-coded by lock type, which is a function of the number of pins in the lock. Labels on each bin and on each container show:

Part number.

The lock type (parent part number).

Containers are about the size of a breadpan. The lot sizes of parts made in the fabrication areas are determined by the number of parts needed to fill an empty container. A full container of parts is placed in the bin area from which it is withdrawn whenever there is an assembly requirement for that lock type. When a container has less than 20 parts left in it, the residual parts are poured into another container holding the same part. The empty container goes to the supplying department for refill rather than going back to the bin.

This system, which eliminates most of the paperwork from these areas of production, was devised 35 years ago by a supervisor. Master Lock gets inventory turns of 10–13 times on the work-in-process (WIP) of the parts on this system, and recently they began to seek ways to improve this.

To improve this system, Master Lock is devising a method, possibly involving move tickets, that will make the supervisor more responsible for keeping stock moving. This will reduce unexpected shortages. An automatic stock rotation system will reduce the losses from obsolescent inventory. Of course, WIP can also be reduced by the simple expedient of removing from the system the containers which do not seem to really be necessary.

Master Lock has no intention of evolving into full automation. They just happen to have a simple but effective method of material control which operates on the same principle as kanban pull signals, and they are doing it in the United States.[3]

---

[3] From a presentation by Pat Murray and Bob Hintz, Master Lock Corporation, at the APICS Repetitive Manufacturing Workshop, June 12, 1981.

4

Planning production to execute a level schedule

The objective of production planning with stockless production is to prepare to execute a level schedule. A level schedule is one that has as even a distribution as possible of material requirements as well as labor requirements. That is, ideally the company would like to have an even distribution of every product made every hour of every day when production is running. Ideals being difficult, they may have to settle for coming as close to this as possible.

Leveling the schedule is only one action required to accomplish this. The physical changes in the plant and equipment must make it possible. Stockless production cannot be achieved without doing both, but a level schedule is a very important part of it. A level schedule assumes that repetitive manufacturing will take place.

One of the more common ways to review the process of production scheduling is in the chronological order by which planning is done, each type of plan becoming more specific up until the plan is executed. The following sequence generally applies to all kinds of production:

Long-range planning. This is a very general plan about overall production levels that goes out well beyond a year. It is based on forecasts. It is used to plan for new plant and equipment, long-range personnel planning, and new product introduction.

Production planning. This is a production plan based on units, labor hours, dollars, or other very coarse units of common measure of production activity level. It is used for refining plans for finance, plant use, budgets, and acquisition of material with a long lead time. Many companies do not develop one in a formal sense, but when they do, it is the plan that corresponds to the top management plans for the business.

Master production schedule. In most companies, this extends over about a one-year planning horizon. It is the major input for planning material requirements, and most companies replan it on a monthly cycle. It may be done on the basis of units, models, types, or material modules. Normally, it becomes more detailed as the plan comes closer to the time of execution. It is commonly prepared in one-week increments of time.

Final assembly schedule. This is the plan for what is really to be assembled day-by-day. It normally has a much shorter planning horizon than the master production schedule. However, in some companies the final assembly schedule is the direct basis of the master production schedule close in on the planning horizon. In others, the final assembly schedule is distinctly different. The master production schedule plans to put material into inventory from which the final assembly schedule can draw as required. It is most commonly prepared in one-day increments of time.

Materials and fabrication schedules. These are typically derived in some way from the master production schedule, and this is always the case by definition when materials requirements planning is used to develop the materials and fabrication schedules.

Prior to stockless production, most companies operated on either a monthly ordering system, or on the basis of Materials Requirements Planning. With monthly ordering, parts schedules are based on forecasts, many of them independently made for each part number. With Materials Requirements Planning they are most generally made by back scheduling due dates for parts based on an explosion of requirements from the Master Production Schedule. If production is repetitive, these schedules are normally converted into daily schedules for the plant floor.

With stockless production, the planning normally follows a similar pattern chronologically, but there are distinct differences. As the ability to execute stockless production improves and lead times become shorter, the planning also becomes simpler. It is easier to present in a logical order rather than chronologically.

DEVELOPING THE LEVEL FINAL ASSEMBLY SCHEDULE

Planning the final assembly schedules requires careful thought because they are the key point at which production is leveled. If fabrication operations are set to supply final assembly by a pull system, then the final assembly schedule is the key that triggers the system. All planning is directed toward being able to develop and maintain level final assembly schedules. If these can be developed, production plans leading to them can be developed.

Very important: Final assembly schedules are planned and replanned. Planning is approximate in the early stages of developing the master production schedule because that is usually a summary in daily buckets of *expected* final assembly schedules. As demand information becomes more firm, the expected final assembly schedules are replanned or adjusted, usually becoming more detailed over time. The final assembly schedules which will really drive the pull system are the last revisions, and these may be developed as little as one day before they are run.

There is a separate final assembly schedule for each final assembly line or final assembly station. Each of these schedules should be level. Then the summation of all of them will also be level. Therefore, what constitutes a level final assembly schedule for just one assembly line?

If a final assembly line is dedicated to producing just one model with a constant bill of material all day every day, that is a level schedule provided the rate of production does not vary. That is the simplest case. If a separate final assembly area could be established for each final product, planning would certainly be simplified. However, this cannot usually be done because the volumes demanded for each model cannot economically justify it, so it is necessary to assemble more than one model in each assembly area. In order to keep the flow of material required at final assembly as level as possible, the assembly lot sizes should be as small as possible. A lot size of one is ideal. Therefore a mixed sequence of all models to be assembled on a given line should be prepared as the schedule.

In order to do this, there must be a summary of the total number of each model (or product) which will be required for a fixed period of time. For example, suppose the fixed period of time is designated to be one month, and for a given month, requirements for final products are:

Model A	400
Model B	300
Model C	200
Model D	100
Total	1,000

A repeating sequence in final assembly that would evenly distribute these four products throughout each day would be

A–B–C–A–B–C–A–B–A–D

This is a cycle that repeats itself every 10 units; and with nice, round numbers for requirements, each product's requirements are proportionally represented in each cycle. Nice round numbers never really occur in actual demand figures, so that problem can be dealt with in two ways:

1. Make small adjustments in the actual demand numbers until proportional numbers are derived. This may be possible when producing either to stock or to order. (For each model, pull a few orders for next period into this one, or push a few orders for late this period into early in the next one.
2. If orders cannot be shifted or adjusted so easily, then set a repeating cycle that comes as close as possible without being horribly complex to execute. Distribute the nonlevel tag-end schedule as evenly as possible throughout the schedule period.

If only a half dozen or so models, each with a constant bill of material, are to be assembled on a given line, the process can be done manually. For products having a huge number of component combination possibilities, as in auto or truck assembly, this must be done by computer. These programs first sort the vehicle orders into various categories, both by structural content and by markets to be served (including promise dates). Then they are sorted into a sequence which approximates a repeating cycle according to priority rules.

After determining the final assembly sequence cycle, the cycle times for each model must be developed. A cycle time is the time between the completion of each model. Returning to the example using the four models A through D, suppose the assembly area worked one shift, from which 420 minutes a day are really used for assembly work, and further assume there are 20 work days in the month:

	$\frac{\text{Monthly requirements}}{\text{Work days per month}}$ = Daily requirements	$\frac{\text{Minutes per day}}{\text{Daily requirements}}$ = Cycle time
Model A	400/20 = 20	420/20 = 21 min.
Model B	300/20 = 15	420/15 = 28 min.
Model C	200/20 = 10	420/10 = 42 min.
Model D	100/20 = 5	420/5 = 84 min.
Overall	1,000/20 = 50	420/50 = 8.4 min.

Cycle time numbers for each model are only divisible by 8.4 if that model is evenly divisible into the total of all models to be run during the day.

Running the mixed model sequence in final assembly means that a unit should be completed every 8.4 minutes, and that among these, a model A unit should be completed every 21 minutes on the average. Not only must the assembly line be physically balanced to run at this rate, but also, since the feeding fabrication operations are to run as nearly in balance as possible, the entire plant is to be balanced to run at this rate. Sometimes that can be done, and sometimes development has not yet progressed to that degree.

The initial stage of final assembly scheduling is to fix the line rates. If final assembly lines are to be balanced and fabrication balanced to run with them, initially fixing the overall line rate is important to allow everyone to plan ahead for the most significant changes in equipment configuration and manning. If equipment is flexible and workers are cross-trained to work at several different positions with skill, the possibility of a plant rebalancing itself to operate at a different overall rate is increased, but that change cannot be made frequently.

On something as complex as an auto assembly line with 1,500 workers and established work station configurations, changing the basic line rate is not simple. It can be done, and sometimes is; but changing a line from, say, 60 units per hour to 45, involves the kind of equipment and manning changes that have to be carefully planned even if operations are flexible, and it is not done every month. What is done much more frequently is to plan for small adjustments in order to operate with different model mixes within the same overall rate, and that is true even for stockless production in its present state. The same kind of situation is to be expected even for operations much less complex than auto assembly if balancing is extended to include fabrication.

What is most often done in Japan to adjust for different overall output requirements is to adjust the length of the planned work day, usually by overtime. The work day will range between 8 and 10 hours, with a 9-hour day being typical.[1] Therefore plants on this system do not plan more than two shifts. With three shifts they would lose some of the necessary ability to react (to model mix changes or to other problems), and any operation which fell behind would have difficulty recovering. Also, the interval between shifts is useful for preventive maintenance. Many operations which must be run around the clock can have the running rate adjusted—for example, polymer extrusion and molding.

If the model mix does not change very much, it might be possible to

[1] The American auto industry has considerable difficulty trying to replicate this practice at present because of existing union agreements, so they adjust overall output in much coarser increments, generally by shutting down a final assembly line for a week from time to time. This on-off scheduling obviously hinders efforts to create better balance between fabrication and assembly.

adjust even an auto assembly line rate somewhere between 55 and 65 units per hour if it is basically manned and equipped for 60 per hour. However, one must be very careful about what this can do to the line balance and material requirement rates if the model mix (or other parts mix) deviates very much from what was originally planned.

Using a simplistic example of the four models A through D, consider what happens if actual demand exceeds the forecast only because of increased demand for model B—not an unusual situation:

Model	*Original forecast*	*Actual*	*Original mix*	*Actual mix*	*Percent change model rate at 55/hour*	*Percent change model rate at 65/hour*
A	400	400	40%	35%	−13%	+18%
B	300	450	30	39	+30	+77
C	200	200	20	17	−13	+18
D	100	100	10	9	−13	+18
Totals	1,000	1,150	100%	100%		

Suppose we had been planning a line rate of 55 per hour (only about 18 hours work, although the time is irrelevant to the point). We could meet the increased demand in two ways: (1) Stay with planned overall rate of 55 per hour and increase the work time by 15 percent, or (2) increase the line rate to 65 per hour and get the production complete in about the same length of time. However, increasing the overall rate has a disastrous effect on the rate of production for model B, which results in a very large increase in the rate of demand of the material for it. If the line rate were kept at 55 per hour, meeting the increased material feed rate and assembly rates for parts unique to model B would be a strain, to say the least. At 65 per hour there is little hope.

In the fabrication operations that feed final assembly, if the same equipment is used to make all the parts for all models, it has a good chance of adjusting to mix changes if the overall rate does not change. The chances are greatly increased by short setup times. If it becomes necessary to change layout, add equipment, and change manning on short notice or ask suppliers to do the same, chances are that the schedule will self-destruct. The same number of total units originally planned should actually be assembled during a planning period. That helps to stabilize the planning for future planning periods as well as to provide the fulcrum for balance of operations while the schedule is running. However, the fabrication work centers supplying parts unique to different models may not be able to see that. Certainly any suppliers who only make parts unique to one model will not.

Some of these considerations require coordinated thought between scheduling and process design. For example, Kawasaki Motorcycle Division in Akashi, Japan, has so far been unable to sequence parts from fabrication into the final assembly line while switching models after every unit. Their assembly rates are not much higher than automotive, but units and workspace are smaller. They sequence parts in fives and run a model mix of five units of each model at a time. Bills of material are fixed, or nearly so, although they run a large variety of models on a single line, and they have found it possible to balance assembly and fabrication material flows in this way. Differences in work station content between models are less a problem than material sorting when staging for the line.

By contrast, the Kawasaki Plant at Lincoln, Nebraska, found it difficult to balance the line if the model mix was sequenced in fives. They build custom orders for large and complex police motorcycles mixed with units having much less part content, but the mix is of only three or four models assembled at a slower rate. It is easier to balance the line by changing models after every unit, just as is true on an auto assembly line.

The lesson in all of this is to carefully develop what models can be assembled and fabricated without making the required changes in tools, work assignments, or material flow impossible. As the total network that will be influenced by the final assembly schedule keeps expanding, these considerations compound. Production planning and control staff need to develop and maintain a set of work rules to guide the development of final assembly schedules; they should work regularly with engineering and plant personnel to keep them updated.

As final assembly schedules are planned and replanned, initially only in rough and finally in detail, the first priority is to set an overall rate that can be met by the existing network for fabrication and assembly. The rate should be such that advantage can be taken of whatever flexibility is currently available for the expected range of model mix variation as it materializes. The final assembly schedules given daily to the assembly lines for execution should have mixes possible within the same rhythm of production throughout a planning period.

MATCHING THE FINAL ASSEMBLY SCHEDULE TO MARKET DEMAND

This is one of the most difficult aspects of stockless production to understand, and several companies have said it is one of the most difficult to do. The amount to assemble of each product or model must be established over a long enough period of time that a level schedule can be developed. A level schedule is one that requires material to be pulled into final assembly in a pattern uniform enough to allow the various elements of production to respond to pull signals. It does not necessarily mean that the usage of every part on an assembly line is identical hour by hour for days on end; it does mean that a given production system equipped with flexible setups and a fixed amount of material in the pipelines can respond.

From the perspective of production planning, stockless production requires development of the ability to synchronize everything from the final assembly schedule. All other schedules are only in preparation for this. Except for final assembly, actual production is executed in response to a pull signal, not a schedule. Schedules received in fabrication are only guidelines for preparation. In planning production for this kind of system,

1. First fix the overall rate of production and an approximate model mix so that everyone can prepare in advance.
2. Within the confines of the overall fixed rate, the model mix can be adjusted, within limits, from that which was first planned.

Of course, if the model mix originally planned can be executed in an identical pattern for days on end, the amount of stock in the pipelines can be squeezed more than otherwise, and everything can be synchronized like clockwork, but the nature of the market demand often does not allow this.

The system thus provides some flexibility to respond to the actual market. The preplanning of final assembly schedules may start three to six months before the final assembly schedules actually run are given daily to the lines. The preplanning is done to a forecast, and the closer the plan comes to the time of execution, the more it is revised, based on a combination of forecast and actual demand. The assembly schedules from which the lines really run should respond to actual demand.

Production planning consists of developing and refining a set of fixed, level rate assembly schedules from which some deviation can be later permitted. The length of time required for production planning depends on how much time is required physically and organizationally for preparation. The amount of time over which the overall production rate is frozen ranges from 5 to 25 working days, but typically this is done in one-month planning periods.

If throughput times are too long, or if a commitment must be made to actually make parts based on a forecast in advance of the final assembly schedule, an operation has reverted to a point where inventory must be held somewhere, and the system must operate, at least in part, by doing calculations of gross-to-net requirements, something not desirable. Example: steel requirements given several months in advance to plan a rolling schedule. In such a case the cumulative planning lead time exceeds the amount of time that can be encompassed by stockless production systems.

From the perspective of the physical capabilities of the production process, the objective is to so reduce the need for inventory and the throughput times of material that all operations can be synchronized in some way to final assembly. The only planning lead times required are those which are necessary in order to have operations track final assembly through the pull system. The length of these planning peri-

LIBRARY ST. MARY'S COLLEGE

ods is a function of the ability to plan and physically prepare for major rate and mix changes. Production networks can be highly developed to do this, but it requires some length of time even for networks in an advanced state of flexible automation.

There is another consideration. Accumulated over short periods of time, the input of demand from the market may exhibit large variances in the mix of models demanded, and even in the total production demanded. The longer that period is, the more likely that variations in market demand will be smoothed, thus giving better selection for keeping the final assembly schedules within the limits which the pull system and existing flexibility can accept.

From the point of view of the market, the length of time the planning schedule is frozen must also not be so long as to be unresponsive. If a schedule is truly frozen, production, in effect, has determined that "the customers have said what they want, and they are going to get it." Most manufacturers serve the kind of market that demands flexibility in product selection up until the last possible moment. Such is the nature of market competition. Therefore deciding the length of the fixed schedule period is something of a compromise. To get a regular rhythm in the planning and communication of changes, an even calendar period is usually selected.

There are really three different kinds of market demand situations to consider:

1. *Make to stock.* Customers order from finished goods inventory, or the finished units are sent into a distribution system.
2. *Make to order.* Production must fulfill the demands of individual customers.
3. *Make for another manufacturer, but not to stock.* The company must respond to the customer's production schedule.

Make to stock. In this case, the determination of how much to make of each model is little different from the determination of how much to make if the schedule is not level. Exhibits 4–1 and 4–2 illustrate a way to do this. The production may be planned directly to a forecast, or those in charge of the finished goods distribution system may determine how much they want to have added to finished goods during the schedule period. In that case, the logistics staff would perform the calculations of the type shown in Exhibit 4–1, and the production planner would see the forecast as an order.

The procedure is very similar to making up an order size with any system. However, the planners should keep in mind that products will be added to the finished goods inventory in small increments—daily in most cases. If the warehouse is some distance from the plant, the frequency of addition will depend on the frequency of transport from the plant. The net effect is that finished goods inventory is reduced.

EXHIBIT 4–1 Example of scheduling model A for a fixed schedule period: Make-to-stock case

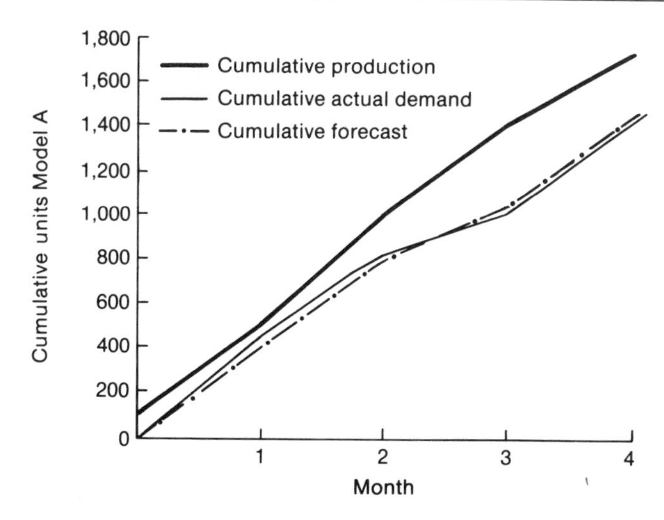

Model A production schedule period development: Make-to-stock case

Month	*1*	*2*	*3*	*4*
Forecast	400	420	370	410
Beginning balance on hand	100	57	118	135
Scheduled production	400	463	352	375
Actual demand	443	402	335	447
Ending balance on hand	57	118	135	63

Target inventory level = 100

Scheduled production rule = Forecast + Target inventory − Ending balance on hand
Example: 463 = 420 + 100 − 57

The amount of stock held as cycle stock from the large order sizes is eliminated. Almost all the stock in finished goods is therefore buffer stock or safety stock. This makes it much easier to relate the level of finished goods carried to the customer service level. The amount of stock should be sufficient to buffer the irregularities in shipments going to customers (and in shipments coming into the warehouse if these are not on a uniform schedule.) The rest is safety stock to allow for deviations between the forecast and actual demand, and time required for production to respond to those deviations.

This kind of planning is not unknown in industry which does not consider itself to be venturing into stockless production. The scheduling of production lines dedicated to the assembly of one model or product has this situation, and scheduling production into finished goods is not generally considered difficult. What uniform production

EXHIBIT 4–2 Comparing effect on finished goods inventory level schedule versus one run a month: example of model A in make-to-stock case

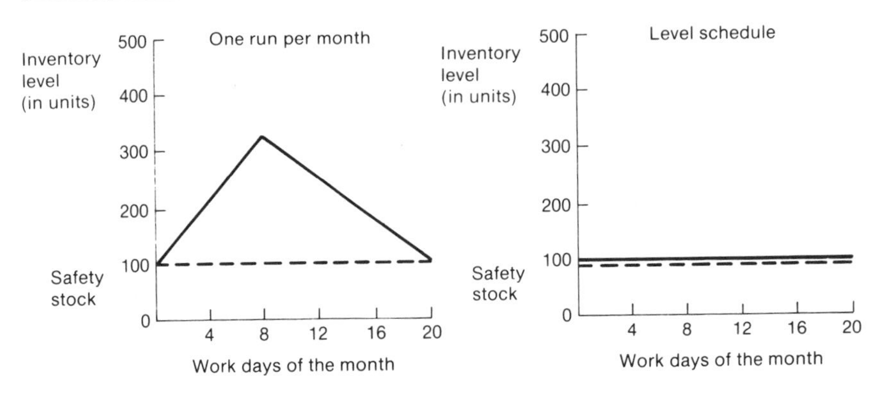

A level schedule removes from finished goods almost all the stock required to allow for intermittent production of different models in long runs. Therefore, with a level schedule, the stock levels in finished goods are almost all safety stock for protection against variations in daily demand rates and deviations of actual demand from forecast. The graph below depicts how the inventory level might look if there were irregular daily shipping patterns *from* finished goods.

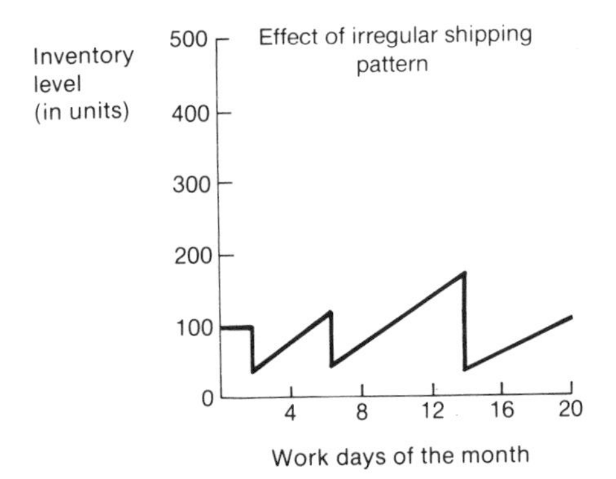

rates are likely to do is concentrate attention on the irregularities in shipping schedules or communications patterns that cause "lumpiness" in demand. The manufacturer provides for this by building up inventory.

The graph in Exhibit 4–1 illustrates that cumulative production tracks sales period by period with stockless production, but if model A in that example incurred a stockout, production would not increase in a sudden spurt to cover it. In fact, such response might not be possible with any production system unless assembly capacity and excess inventory were ready, and that is not what is wanted. Rather, we are concerned with how long it will take to change production all through the network to support this increase in production, thinking of the entire production process as something like assembly.

The degree to which production can respond to changes in demand coming from the finished goods system is a function again of the flexibility developed in the physical production system all the way down the network, and of the planning lead time required to execute a major change. If the changes are not great, it should be possible to absorb them in changes in the mix in final assembly schedules, as described in the previous section. If the daily assembly rates are held constant for periods shorter than a month and mix changes can take place at intervals of less than 10 days, somewhat larger changes are possible. A change extreme enough to require major changes in capacity or equipment used will require a still longer planning lead time, but that is really no different from what prevails in most manufacturing anyway. If the production process has been revised for more flexibility, it may be a faster and quicker reaction time. In fact, one of the reasons for desiring stockless production is sometimes to take advantage of the ability to be more flexible, not just to reduce production costs.

In designing the production planning and control system, one of the vital questions in the make-to-stock case is how often to replan, based on a new batch of "orders" from the finished goods system. The shorter this period is, the closer production can track actual demand changes.

Stockless production can be used for production of seasonal items. The planned assembly rates and model mixes are changed in blocks according to the season. It does not do very much to solve the old problem of whether to run production based on forecasts in the off season, but it may help a little. If the production system has shorter throughput times and shorter production planning lead times, it should be possible to adopt schedules that are closer to "chase" strategies, but risks are certainly not eliminated without being able to run a perfect chase schedule. (Kawasaki USA built a lot of snowmobiles in anticipation of snow that never came.)

Make to order. This case is generally more difficult to deal with. The scheduler needs to have a backlog of customers' orders of sufficient mix to develop final assembly schedules within the rate and mix parameters set by a plan probably developed mostly on forecast. The scheduling is also made more difficult because products built to order are also the ones which have more model variations and customer options, which creates more complexity in establishing the mix range parameters when planning for a level schedule. A brief overview of the production planning process used to handle this is presented in Exhibit 4–3.

EXHIBIT 4–3 Production planning for stockless production

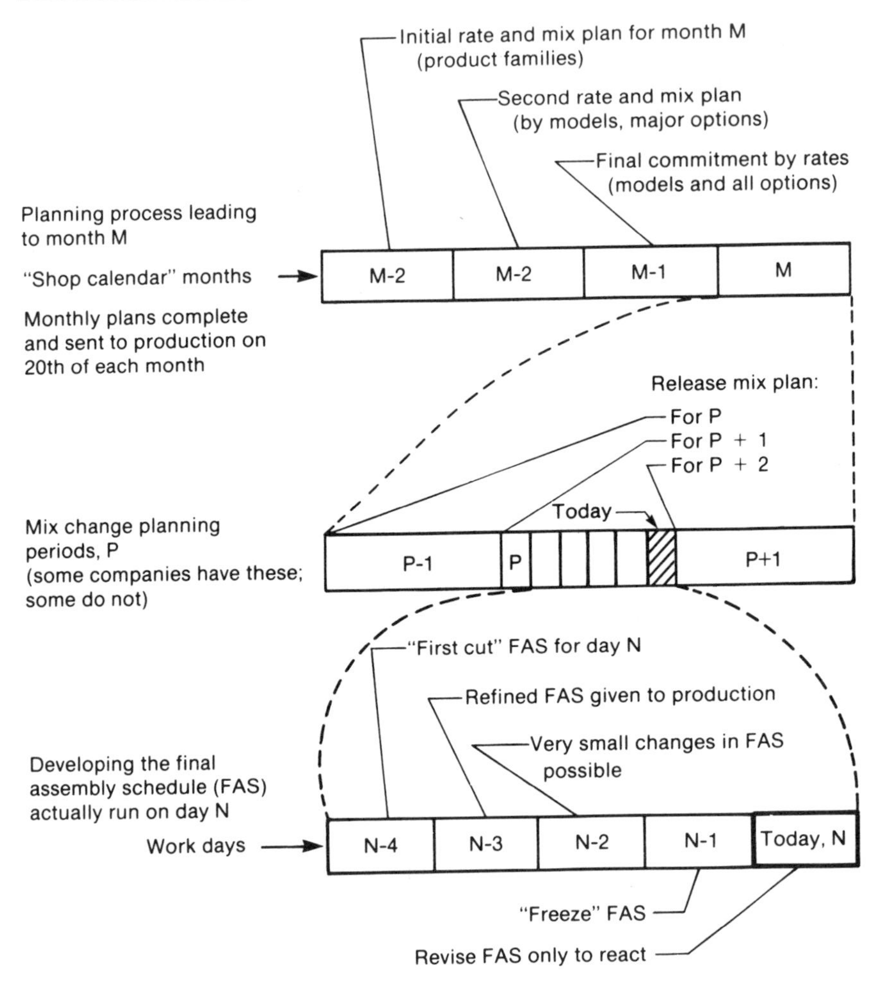

This hypothetical example is very similar to the system used by several companies but not identical to any of them. Lengths of planning periods and horizons vary from industry to industry. (Auto companies are very similar.)

The more days of actual orders that are in the backlog, the better the selection of orders to fill daily schedules which adhere to the mix parameters within the capability of the total production network, including suppliers. Likewise, the more flexible the total production system being controlled, the fewer the restrictions in selecting orders for each day's assembly. However, having a large order backlog for

EXHIBIT 4–4 Consumption of make-to-stock orders in a final assembly schedule (FAS)

With 10 days of orders on hand:

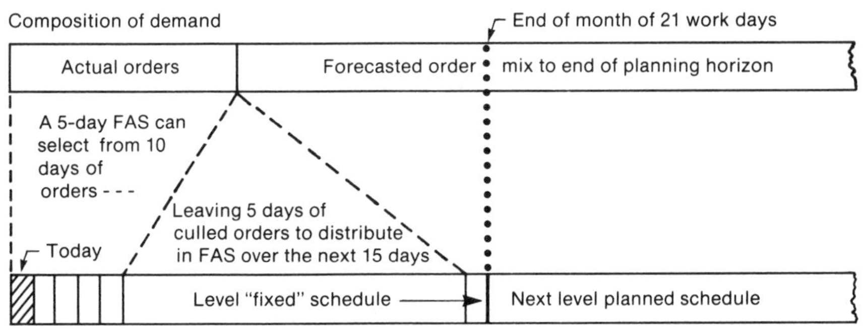

With 15 days of orders on hand:

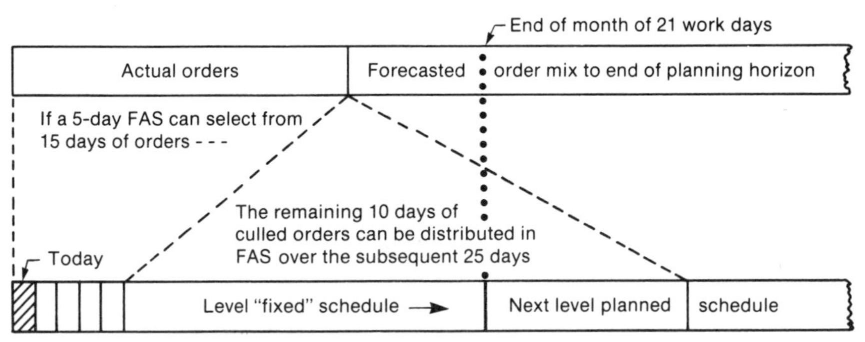

The above example makes some simplifying assumptions to illustrate the point:

Five days are consumed to develop the daily FAS.

Order promise dates have a horizon the same length as the number of days' on hand—often not the case.

Orders in the order backlog are firm orders—sometimes not the case.

The overall assembly rates and basic mix are leveled firm over monthly planning periods.

planning purposes also means having orders in hand that are not being sent to the customers, and from the customers' viewpoint, orders held for sorting still represent lead time to them just as if material were stuck in a queue in a job shop. Once again, the ultimate solution is physical revision to have flexibility and short throughput times, not superplanning.

The same kinds of problems that occur in make-to-order may also occur in make-to-stock if the product mix is complex. In such a case, the interaction between the finished goods distribution system and production may be almost like customers sending custom orders.

Some companies build partly to stock and partly to order. They can intersperse built-to-stock units with those built-to-order. They may even have an agreement between marketing and production to accept some extra units into finished goods stock if these are useful as fillers in creating a level schedule. There is also the temptation for the completely make-to-order company to add a few of the most popular models built ahead in the schedule if that will help preserve a producible mix—a dangerous practice because it starts to erode the discipline of matching product assembly to the market. (Toyota says they once had such an agreement with their marketing arm, but in all the years they have operated this way, they have never built more than 1 percent "John Doe" units in any month. They do not want to allow this because it not only defeats what they are trying to accomplish with stockless production, but it is also a dangerous business risk if it gets out of hand.)

In the case where no make-to-stock can be permitted, the schedule development is a process of transforming the pattern of order arrivals into a level pattern for final assembly schedules without any stock units to make the task easier. This gets into the heart of what is usually a rather complex activity of order entry in a company. If the transformation of orders into schedules is to take place efficiently, it should start as soon as possible after orders arrive, but this depends on the sales policies and trade customs affecting the arriving orders. For example:

1. How much time is needed for credit checks or confirmations and the resulting turn-down ratio?
2. How much time is required to put the order into a format for schedule planning purposes? Must the order be checked for consistency and completeness? Must it be checked for technical feasibility or material availability?
3. What are the policies on changes and cancellations? What percentage of the backlogged orders are stable?
4. What flexibility is there in assigning due dates or in working to them?

5. How many variations are there in the bill of materials for the product? A variation is defined as a combination of specifications determined by the nature of the order, but the combination is not specified in detail by the customer. An example comes from the auto industry in which exported autos must be equipped to meet the regulations of the country of destination. A large number of parts and subassemblies may be affected, but the variation is a fixed package for the given purpose. The same condition applies to electronic equipment or appliances which are exported, or to industrial products which have different standard specifications for different industries.
6. How many options are permitted and what is their nature? Options are components and subassemblies specified by the customer. If customers are allowed to specify options only in packages and not in any feasible combination, that obviously helps to assist in selecting orders which fit within the mix parameters allowable by a given level schedule. Also some options may be simple "hang ons" while others create a great ripple effect by the total number of different part and subassembly combinations which must be allowed in the mix in order to accommodate them.

These policies, and others like them, affect the stability of the order backlog, and also the number of orders which are really available and on which schedule development is based. A 60-day backlog of orders means little if only 10 days' orders are in a condition for assignment to a schedule.

Much has been made of the impossibility of combining product options with a stockless production level schedule. It is how options are handled rather than the product complexity itself that counts. Because of the countries of export served, a Toyota assembly plant has many more product variations to service than do American auto assembly plants, but standard packages in a predictable sales pattern can be worked into an allowable schedule mix. What is troublesome is having a large number of options which customers can specify in unique combinations which work their way deep into the bill of material. Toyota, however, restricts customers to ordering option packages, then tries to turn this into a marketing advantage through low cost rather than have it be a marketing disadvantage.

The pattern of order arrivals is also important. If sales procedures are such that order arrivals peak at the end of the month, the schedule development process must take that into account. If the surges in orders come to regular intervals, it may be possible to build the planning process around it, sorting orders for many consecutive days at the peak time. However, many days' orders on hand at once means that some of those orders will have a delay time before delivery. Then the schedulers also have to figure out how to get through low ebb in the order backlog. Both customer service and production effectiveness seem better served by a stable order backlog unless orders coming in waves are a natural part of the market being served.

The same is true if orders for particular models arrive in surges, causing the order backlog to go off the chart in model mix selection. Examples: fleet vehicle orders, government orders for electronic equipment, appliance orders for large housing projects. These will not fit into the scheduled mix, so the delivery lead time on the orders must be long enough that the production network can plan for the changes they require—perhaps not much different from the delivery times such customers would see without stockless production; but again, if the total network has attained a high degree of flexibility, the response time for such orders may be improved in total lead time to the customer.

Make for another manufacturer. For products shipped directly to another manufacturer, it is best if both companies are on stockless production with the schedules tightly linked. However, that may not develop quickly. If the supplying company runs by a level schedule, but its customers do not, there is no recourse but to buffer the demands of the customer from the level schedule by building up finished goods inventory. Almost all supplier companies practicing stockless production have at least a few such customers. It is necessary to build up the inventory to meet a bulk shipping requirement. It is a good idea to have a policy established with such customers concerning the size of the "parts bank" which is held in finished goods for them. The policy should have the objective of precluding the customer making demands on production that would wreck the level schedule. Aside from the need to protect the level schedule, there is really very little different about this policy from what might be set up in the absence of stockless production.

Even if all of a supplier company's industrial customers use their outputs in long production runs lasting weeks, a supplier company still may run a mixed model level assembly schedule itself. The internal benefits in improving operations are the reason for doing so. However, this company will certainly have incentive to try to modify its equipment to handle very large swings in model mix.

Even if the supplier company serves industrial customers who are themselves running a level schedule, they will wish to handle as wide a range of model mix as possible and perhaps even be able to alter output levels frequently. The reason goes back to the variations in model rates in the customer plants. A supplier who provides parts for only half the models in the mix of a level schedule can easily be confronted with ±15–20 percent daily swings in volumes within the mix demanded. Unless measures are taken to control this, this variation will amplify as it cascades down through several stages of production and through tiers of suppliers.

The best way to control such variations begins with selection of the parts a supplier will feed to a given schedule of each stockless production customer. If the same supplier plant supplies variations of the same part for all of the models in a customer's level schedule, the supplier will at least see no more variation than is allowed in the customer's own schedule. The less chance the supplier has of being able to track the customer through a pull system, the more likely that the operations between the two companies must be buffered by ups and downs in an inventory held somewhere between them.

Experience with stockless production is that the amplitude of rate and mix changes increases as it goes through each stage of production further removed from the level final assembly schedule for the end products themselves. Therefore, suppliers need to design equipment, plants, and scheduling system so that they can be made more reactive. Smaller aggregations of systems and equipment can generally be made more flexible and reactive, so suppliers should try to avoid having too many levels of process interaction of their own.

Many companies and plants serve multiple sources of demand: to order, to stock, and to supply other manufacturers. Some industries have special problems in coordinating orders between them. For example, computer companies generally want to coordinate the shipment of mainframe and all peripheral equipment to arrive at a customer's site at the same time, so the subassemblies never enter the "customer" plant or become a direct part of final assembly scheduling. If the amount of market complexity served is too great, the difficulty in creating a level schedule becomes overwhelming. This is a strong argument for carefully thinking through what is necessary in order to have a focused factory for stockless production purposes.

Finally, problems in converting demand to a final assembly schedule that can initiate a pull system are problems that add to the lead time to the customer. Like much else about stockless production, it is not hard to understand how to prepare a level schedule, but it is hard to simplify things in order to accomplish it.

RESCHEDULING THE FINAL ASSEMBLY SCHEDULE

In a sense, no factory is ever scheduled—it is only rescheduled. A final assembly schedule is intended to be executed to the unit, but sometimes events preclude that from happening. Orders are cancelled. Trucks wreck. Dies break unexpectedly despite the best of preventive maintenance programs. Therefore, it is sometimes necessary to rework an unfinished block of leveled final assembly schedule.

This is an everyday occurrence in the final assembly schedule in auto plants. If a car body is found to have defective sheet metal upon entering the paint shop, that unit must be pulled out of sequence for the rest of the assembly. In fact the planned final assembly sequence is not set until three to five days before the schedule is run, and minor adjustments occur after that if there are problems. (Recall that leveling the schedule means running to a *planned* level schedule. It does not mean that the schedule is executed to the unit at all times.) Sometimes, the schedule gets a little behind, and people must work overtime. Executing a level schedule means finishing the total number of units originally scheduled, with the original model mix pattern adhered to as closely as possible. If demand suddenly drops, the number of hours worked may be decreased and a slightly revised mix pattern adopted, but in no case should the total number of units left incomplete at the end of one schedule period be significant enough to require revision of the basic plan for the subsequent period. That will degrade the system if it is permitted.

The purpose of rescheduling is to finish the units planned for a week or for a fixed schedule period despite some obstacles which have arisen. An example comes from the Kawasaki plant at Lincoln, Nebraska. The plant maintains a two-to-three-day buffer of parts inbound from Japan, and those parts have about a six-week lead time and a long transport time. Once in a while a part for one model does not come in. Then that model has to be dropped from the assembly sequence until it is possible to run it again. In that case the production control manager manually recalculates a new assembly sequence for the day. As soon as the part comes (usually within 24 hours), the manager revises the assembly sequence again to increase the frequency of assembly of the model which was omitted from production for one day. The rescheduling process for one line is done manually in about a half hour. However, this is not desirable practice. The system should not be set up only to manage a crisis; it should also facilitate a detailed study in how to prevent recurrence of the crisis.

THE PRODUCTION PLANNING PROCESS LEADING TO THE FINAL ASSEMBLY SCHEDULE

The materials planning system evolves into the final assembly schedule over time. It is difficult to describe such a system without an example, so the one used will be typical of the Japanese auto industry. Each stockless production company in any industry has a somewhat similar pattern of planning. It makes a difference in these systems if a company makes to order, makes to stock, or supplies another manufacturer, but the differences are not great beyond the point where actual orders for stock or for customers must be converted to schedule. All planning develops from a forecast in the early stages. (See Exhibit 4–5.)

EXHIBIT 4–5 Summary of typical production planning schedule for stockless production

Planning stage	*Purpose*	*Planning horizon*	*Planning cycle*	*Format*	*Developed From*
Long-range plan	Plan new products and models. Plan facilities. Plan systems changes.	3–5 yrs.	6 months to 1 year	Indefinite	Corporate forecasts and policies
Production plan	Plan tools and materials. Supplier planning. Total manpower.	1 yr.	1 month	Summarized by model or family in monthly buckets	Long range plan and more detailed forecasts
Master production schedule	Develop data over time to evolve into final assembly schedule. Make major manpower allocations.	3–6 months*	1 month*	Summary of expected final assembly schedule in daily buckets	Production plan and more detailed forecasts and actual orders. Sometimes from finished goods system
Planned fabrication schedules	Plan ahead for the support required to feed the planned final assembly schedules: Department planning, Supplier planning, Transport planning, Manning, etc.	3–6 months	1 month	Quantities, cycle times and WIP levels for each identical day of the level time blocks of the final assembly schedule	Exploded from the master production schedule
Final assembly schedule	1. Match market demand rate. 2. Trigger the pull systems of control. 3. Level the total production process.	2–5 days	Daily**	A sequence of end units to be assembled each day. Repeating sequences, cycle times and data need to run level schedule	Refined from feedback based on the master production schedule and planned fabrication schedules

* There may be shorter times and horizons for mix change plans for those companies which have them.

** The horizon and cycle may be somewhat longer if a company only makes to stock. It depends on the planning cycle of the finished goods system.

Long-range planning does not follow a similar format from company to company. It typically is done for planning new products and facilities.

The *production plan* is typically the earliest plan for material. It usually has a one-year planning horizon and is often little more than massaged forecasts used for ordering long lead-time raw material or tooling. It is also the basis of planning with suppliers. Agreements with suppliers are typically replanned about every six months, based on updated forecasts.

The *master production schedule* is a summary in daily buckets of the numbers of each model expected to be built in the final assembly schedule. It can be thought of as a series of worksheets which become more detailed until they evolve into the final assembly schedule. It typically has a three-to-six month planning horizon:

—It is usually planned in 10-day blocks, or buckets, in which every day's schedule is identical—or nearly enough identical that each day can be run without any concern for having to revise manning or tooling plans for each work center.

—It is usually replanned on a monthly planning cycle. The first month is usually frozen and in progress. More precisely, the 10-day block currently in progress is frozen, and so is the one following it, and a third one may be frozen, depending on time of the month. The frozen portion of the schedule is a daily summary detailed by end item.

—The second month consists of two-to-three more 10-day blocks in planning, daily summaries by model of the proposed final assembly schedules.

—The third and following months are generally an initial plan for final assembly typically detailed by models or by model families as summaries of daily assembly quantities.

—Many planners develop the master schedule manually. As more and more detail is added, a computer becomes necessary. This is true if planners are trying to estimate many different assembly cycle times for different models.

—In most companies, the timing of new model introductions, or trial runs, and the timing of engineering changes are planned as part of the master production schedule.

Planned fabrication schedules are developed by exploding the master production schedules using the bills of materials. Since every day's master production schedule is identical for each day of a 10-day block, the fabrication schedules are the same every day within each 10-day block. Therefore planning does not require time offsetting as with MRP. (The key characteristic of

MRP is backscheduling fabrication and purchasing schedules from due dates. Product explosions are used with other forms of materials planning.)

—What results from this is a set of fabrication schedules, one for each identical day in the 10-day planned blocks, or for each day of the planned months of the master production schedule.

—The purpose of these schedules is to allow fabrication, subassembly, and supplier supervisors to have advance warning about the scope of schedules to be run. An important piece of information is the expected cycle rates planned for production. This allows preplanning of the workplace, manning, and tooling organization required to balance the operations, move the material, and perform preventive maintenance.

—Capacity planning generally comes from the department managers reviewing these schedules and determining if they can organize their department to run the schedule. This generally has a purpose somewhat different from a job shop where capacity planning is a comparison of projected loads and work center capabilities. The advanced planning for stockless production should provide for capacity in excess of what is required well in advance of this state of planning, thus allowing open machine time for maintenance and catch-up. (Department heads should attend regular scheduling meetings to discuss their department's state of readiness to operate by a tentative planned schedule.)

—The planners may need to revise the master schedule if feedback from managers indicates difficulty in running the schedule. But the goal is flexibility—to reduce the frequency of revision.

—These schedules are also sent to suppliers and discussed with them. The supplier schedules in Japan must be developed into a set of daily expected quantities and arrival times for each part.

—Likewise, the planned delivery quantities and times must be sent to the trucking companies for their planning. The transport capacity is also important; it may be necessary to increase or decrease the frequency of deliveries between plants and to revise truck schedules to avoid bunched arrivals of trucks at the same plant.

—These delivery schedules also include delivery schedules on a frequent basis from most raw material suppliers. Steel is typically delivered as frequently as twice a week to once a day.

—In a similar way, the pattern of material handling expected inside the plants must be considered to be sure that material handlers have a level load, and that material transfers do not choke in their own congestion at various times and places.

—Following this planning, a very important activity is estimating work-in-process inventory levels needed to support the plan and the volumes in the pipelines between plants and between work centers within plants. This is sometimes called planning the depth of the process, an analogy to inventory as the depth of water in a flow. This may be done by approximation in aggregates at higher company levels and become more detailed as planning comes closer to each plant.

—The depth of the process is so closely tied to the capability of each department to operate efficiently that the production control department can only issue proposed inventory levels. Department managers and foremen must adjust the WIP levels for parts to allow for the current status of problems or improvements in their departments.

—The planned fabrication schedule only advises departments of the impact of the planned Final Assembly Schedule on each part. Actual fabrication takes place in response to the pull system coming from final assembly. Schedulers must keep in mind that the production departments will use these schedules to prepare for:

Material handling
Tool schedules
Preventive maintenance
Quality check routines
Equipment configurations and layout
Manning
Ongoing improvement projects

Final Assembly Schedules are developed as the final stage of planning. The production control department gives the schedules to the final assembly areas. Actually executing the schedules so that demand for material and maintenance is uniform throughout the production period is not very easy, so Production Control must constantly monitor performance to the schedule and take corrective action as required. The responsibility for producing in synchronization with the final assembly schedule falls to the department heads and foremen. Production Control only helps by adjusting the schedules, if necessary, to solve problems. Again, their objective is to make the adjustment meet the goal of fulfilling the planned schedule with as little deviation as possible.

APPENDIX

The story below briefly describes the production planning and control system of a nonautomotive American repetitive manufacturer. Although the system is described as a pull system, readers will detect that this is a matter of degree. In contrast with many production planning and control systems, this one drives more of the action from the final assembly schedule, but it does not

1. Exclusively make commitments to production from pull signals derived from final assembly.
2. Restrict the volume of inventory in the pipelines so that production responds to signals, not gross-to-net inventory calculations.

It is remarkably similar to the kinds of systems many Japanese stockless production companies had in use just *prior* to their stockless production programs.

(Author's note.)

A BASIC PULL SYSTEM AT HOOVER COMPANY[2]

The Hoover Company manufactures a full line of vacuum cleaners in two plants at North Canton, Ohio, having a combined work force of 2,200 hourly employees and about 1¼ million square feet of floor space. These plants operate effectively using a basic pull system, developed over a number of years, for production scheduling a material flow control of 360 different models and 29,000 part numbers. They have looked at other more publicized American systems but have stayed with their own system because they feel it is more cost effective than the highly sophisticated alternatives.

A monthly assembly schedule of the various models of vacuum cleaners is prepared to meet the needs of actual orders and requirements of the distribution system. This is adjusted as needed to reflect the existing plant load, capacity limitations, and other changing conditions. The schedule is a summary in daily buckets of what is expected to be in the final assembly schedule each month. It is used as a master schedule.

The monthly master schedule is exploded to provide the total gross requirements of all parts and subassemblies to be purchased or manufactured each month. It is basically a single-level explosion. Material schedulers and manufacturing department supervisors receive a copy of this gross requirements report. It lists the requirements for each part needed to support the monthly assembly schedule. The report shows the end-item model for which each part is designated in assembly. Gross-to-net requirement calculations are also done by the schedulers and supervisors. These calculations consider stocks on hand.

In addition, everyone concerned receives a copy of the final assembly schedule. This shows, for each assembly line, the planned assembly rate of each model by day and the length of the assembly run for each. By referring to the final assembly schedule, material schedulers and supervisors can see the days during the month when each part

[2] Description courtesy of Donald J. Geis and the Hoover Company.

will be needed. They determine when to schedule supplier deliveries or internal parts manufacture by offsetting from final assembly dates, using lead times which they keep themselves.

Lot sizes for purchased and manufactured items are also determined by material schedulers and production supervisors. Empirical lot-sizing rules prevail, with consideration given to setup and purchase costs, storage space required, and inventory investment costs. Delivery of purchased and manufactured parts is keyed to the final assembly schedule. Raw material for parts fabrication is scheduled for delivery, using more general guidelines. Examples: For high volume material, deliver a few days' worth at a time; lower volume material will be delivered in several weeks' or a month's quantity.

It is sometimes necessary to reschedule supplier deliveries or in-house fabrication to compensate for situations such as delivery failures, component or material quality problems, or marketing requirements. This rescheduling is done informally; personal communication is the key to doing this successfully. In fact, personal communication is an important element to making the whole production scheduling and material control system work effectively. People know each other well. They are accustomed to each other's decision rules and have learned to anticipate problems so as to minimize expediting. There is a cooperative attitude among Hoover's people; they work well together to gain high schedule achievement with low inventory investment.

These plants run with 10 to 12 annual turns of manufacturing inventory (raw material and work in process; finished goods are not included, as they are moved daily to the distribution system.) These plants operate without stockrooms, without shop orders, job packets, or shop floor expediters. Hoover has affiliate plants in other countries manufacturing essentially the same types of products, using more formal control systems. The affiliate plants generally do not achieve as good results in manufacturing productivity and inventory turnover.

Hoover's U.S. system is not designed to be rigid and formal. It is effective because it is flexible and simple to operate. It is an example of what Americans do in repetitive manufacturing as a natural matter of course. Hoover continually looks at new systems and techniques, insisting that any changes must prove as cost-effective as the present system and be simple to operate from the perspective of its use on the shop floor.

5

Converting the plant for flow and flexibility

Planning and control systems are only one part of stockless production. They cannot be implemented unless the methods of production are designed to permit them. Much of the productivity benefit from stockless production comes from the revision of manufacturing processes. The degree to which this is true is hard to quantify, but managers familiar with the methods generally estimate that it is in the range of 15–20 percent of the benefit coming from material planning and control. The rest comes from upgrading the manufacturing process and from the increased effectiveness of workers as a result.

There are five areas of manufacturing process improvement, but only the first two will be discussed in this chapter (the rest later):

1. Reduction of setup times: increased flexibility of production.
2. Revised plant layout and attention to housekeeping to streamline flows of tools and material and also to promote the visibility of problems.
3. Preventive maintenance and other similar programs to anticipate problems.
4. Balancing of operation cycle times to tightly link operations together.
5. Developing the system into full automation.

A plant can be designed new with as much flexible automation as engineering can muster, but it is not possible for engineers to anticipate everything that will be required. Every new plant must be refined and debugged, an evolutionary action process.

In existing plants, the equipment and layout evolve into the condition necessary for flexible flow. The intention is to make changes, learn from the results, and make more changes, all the while removing as much "black art" as possible from the production processes. It is not possible for a few engineers to do this alone. Plant flexibility and flow rates increase not only because equipment and layout are modified, but because the operators develop themselves to use more skill with less effort.

Almost every plant not designed for stockless production starts from a muddle. Everyone starts developing the discipline for conversion by giving detailed attention to reduction of setup times and to revision of the plant layout.

SETUP-TIME REDUCTION

In stockless production the definition of setup time for any process is the elapsed downtime between the last production piece of part A and the first good production piece of part B. Setup is not complete if the process is still running scrap and trying to make production. Likewise, setup time is *not* the total labor time required for the setup. It is certainly desirable to reduce that also, but production flexibility is increased by reducing process downtime—last piece to first piece.

This is essential for stockless production, and it usually requires modification of existing equipment. Buying new equipment for the purpose is not only costly, but also may not result in the kind of change wanted. It is better to first gain experience in what is wanted by studying present equipment. This is a study-and-action procedure, here presented in several steps, though there is no simple by-the-numbers method for going about it.

A program to reduce setup times is very much like any other program of industrial engineering methods study. Since much of the work of changing the setup on any equipment is a matter of making mechanical changes and revising the layout of material and items around it, much can be done to reduce setup times without resorting to advanced engineering. Technicians can accomplish a great deal, but it takes time before the know-how for it accumulates. Every company has its own assortment of equipment and its own set of production tasks to be accomplished on that equipment. People who are thoroughly familiar both with the equipment and what it is supposed to do are best suited to figure out how to make setup time reductions.

Programs start with goals of cutting setup times in half, then down to half an hour, and so on, but the major accomplishment is to reduce equipment setup times to 10 minutes or less. That is called a single-digit setup. After considerable work and study, even large presses have single-digit setup times.

When possible, this goal is extended to achieve setups in less than a minute. Then every move in the setup has to be the right one, with none of them repeated and no hesitation. A single worker may be able to perform an easy setup in seconds with one hand; hence it is called a one-touch setup, possible for simple, lightweight equipment. However, the term is used to denote setups which are flawlessly executed in times somewhat over a minute as a routine operating procedure.

The ultimate is for the machine to automatically set up itself, sometimes true for programmed equipment and for some that is not programmed. (Tokai Rika Company has a small press rigged so that it is capable of changing lightweight dies with every stroke of the press.)

With dedication to the task, companies have reduced setup times of common types of presses and machine tools by 80 percent in the course of a year, with plenty of potential left for further setup time reduction after that. An outstanding setup time reduction program is diagrammed in Exhibit 5–1. Company A had a number of people thoroughly familiar with ideas for reducing setup times on die cast machines when they started.

Those who have never before considered a program to reduce setup times frequently look with disbelief at the setup times listed by companies well advanced in the practice. However, setup times are reduced in increments by a mixture of good industrial engineering methods analysis, mechanics, a little instrumentation, and common sense. The basic line of attack is outlined in Exhibit 5–2.

Many of the procedural ideas to make changes are simple ones. Frequently people say, "Oh, I've seen that!" or "We've used that, too!" Many of these are unseen because we become accepting of existing methods of operation until they are not seriously questioned, and this is especially true of setup times. Ideas that we do have often go a long time before implementation because there is no program to continuously put small improvements into effect. On the plant floor a major element of stockless production is to constantly rethink what is being done and get around to it. Getting in mind what we are trying to do and why is more difficult than concocting many of the ideas for physical change.

It puts a different light on the study of setup procedures to resolve that people in the plant are going to set up multiple times daily as a regular thing. After all, if the plant is going to become more automated, the major work on the floor will be setup and not running parts. Therefore, operators must be able to make setups in standard ways as a routine operation. Then we begin to concentrate attention on setup

EXHIBIT 5–1 Company A die cast setup project

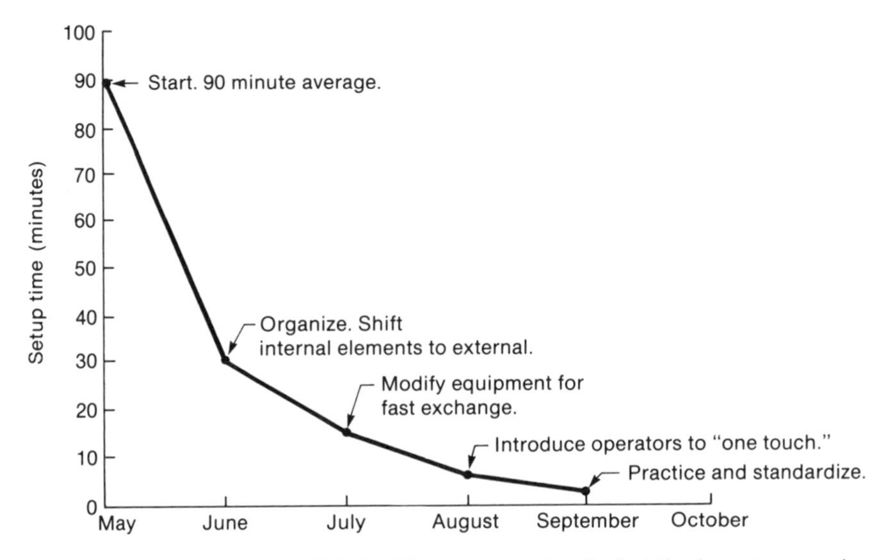

Various companies give slightly different accounts of what the key steps are in reducing setup times, but many of them advocate the three key steps designated by Shigeo Shingo as most important:

1. Displacing internal setup activities to external—doing as much as possible with the machine running.
2. Modifying equipment for rapid exchange during internal setup.
3. Eliminating adjustments.

operations just as we have traditionally concentrated on improving the running operations.

1. Examine and reexamine what a setup is to accomplish

Look at the future of the production process. Effort spent reducing setup time on existing equipment can be wasted without it being guided from an overall viewpoint—the "big picture." The company may be about to drop one of the products being made. It may be obtaining new equipment. Engineering may be doing a redesign of the product that will affect both the problems and the opportunities for setup reduction. Making sure that those working on the details of setup time reduction are aware of these things is the obligation of upper-level management and a normal part of their duties. However, everyone in manufacturing for a few years has seen major projects evaporate because the underlying assumptions for them evaporated.

EXHIBIT 5–2 Overview of setup-time reduction methods

1. Examine and reexamine what a setup is to accomplish.
   a. Review the "big picture"—planned product and equipment changes.
   b. Is each process for which setup is now required really necessary?
   c. Establish what parts or range of tasks a specific machine setup pattern must cover, which depends on the processes it is to supply and how. That depends on establishing the flow path of material from final assembly back to it.
2. Study and document how current setups are accomplished.
   a. Detail as much as possible.
   b. Generate ideas. Direct ideas and improvements based on the Pareto principle.
   c. Base further analysis on the results of improvements made.
3. Reduce machine downtime as much as possible.
   a. Transfer setup activity with the machine stopped (internal setup) to activity performed with the machine running (external setup).
      (1) Revise procedures and layout to stage for internal setup.
      (2) Eliminate lifting where possible.
      (3) Make it a one-operator operation if possible—standard, routine methods.
      (4) Develop a team procedure where necessary.
      (5) Modify the machine, attachments, and tools to perform more operations with the machine running, fast operations with the machine stopped.
   b. Modify equipment to promote rapid internal setup.
      (1) Cartridge-type single-position connections.
      (2) Color coding, multiple-connection plugs, and so forth.
   c. Eliminate as much adjustment time as possible.
      (1) Be sure the basic fit of all items does not require unnecessary positioning adjustment.
      (2) Instrument equipment as necessary to permit establishing setup conditions without trial and error.
      (3) Determine, record, and mark setup conditions so that they can be easily reproduced time after time. For example, record and post feeds, temperatures, and so forth. Use step-function adjustments, similar to radio push-button tuning.
4. Organize, kit, and sequence material for assembly and subassembly to enable rapidly changing from assembly of one item to assembly of another—assembly of a mixed-model sequence, if possible.
5. Practice and refine the setup procedures.

This kind of guidance is often not very easy to give. It helps if management reviews what parts and products are to be processed in a given plant—a program to "focus" the factories.

What range of setups on a given machine? Begin at final assembly and work backwards to determine what each machine will be assigned to do. In the process of streamlining the flow so that one part is made only in one place, the mission of each piece of equipment will usually be narrowed so that the range of setup tasks required is simplified.

Eventually one comes around to the essential questions: How many different dies on a specific press? How many different camshafts to go on a cam grinder, and how significant are the differences in configuration of each camshaft? If we are using a screw machine for simple cutoff operations, will equipment simpler for setup do as well, or can the limited range of setups be done quickly enough on the screw machine itself?

This review can be as fundamental and technical as necessary, but if it becomes a project to do basic research on the material conversion processes of production, the purpose is defeated. Try to keep the changes in equipment selection restricted to equipment which is available and relatively easy to understand at first. Make simple changes to modify what is available to do the job.

Example: Suppose that a single large extruder is used for parts of two different colors. Changing colors requires either removing the screw for cleaning or a purge between colors, and either one takes time and creates scrap. If it is possible to use two extruders, one for each color, the study of setup time reduction can be confined to how to rapidly exchange dies. Of course, this may require capital cost if one concludes that the solution is to go out and buy two different extruders. An imaginative program of equipment selection and exchange over time may accomplish this without breaking the capital budget.

Example: For many products, such as motorcycles and appliances, each final assembly unit may require 15–25 miscellaneous parts all painted black for each model. If each part were to be painted separately, that would require a long sequence of setups, each of which involved only changing the part being hung on a hook and the spray pattern used. Since most of these parts are small, generally several different parts are hung on hooks in groups. However, if the assortment of parts hung in a group does not match the set which is required for assembly of a specific model, the setup problem in paint is to keep enough parts of all types on the output side so assembly has enough parts of each type at the time needed. That obviously requires some stock on the inbound side of paint and the outbound side of paint to cover the setup changes, and the paint schedule which keeps the line supplied may be hard to control.

The preferred solution is to modify this paint operation so that it can accept matched sets of all the parts for any model. Then paint the matched sets in approximately the sequence to be used in final assembly even if it runs a mixed-model schedule. If possible, have the output from paint be in the groups which will be taken to the correct line stations for use. That requires a lot of attention to the parts positioning and selection when they are hung, but the parts need to be sorted out in this way at some point in production, and the earlier that can be done and kept that way the better. The level schedule is thus driven further back into the process, and the departments feeding parts to paint also see a level schedule. What starts out perceived as a setup problem turns into a basic revision of the procedures and process.

The point of these examples is that a plant-wide program to reduce setup times becomes entangled with all kinds of issues and problems, including the obvious one of whether all the variety of parts made is really needed. It can involve value engineering, the basic technology of the processes, and so on. At some point people are likely to conclude that the total setup reduction program should be delayed until the big problems of process redesign are solved, so the advice part of this is the following:

1. Get a basic idea of what the eventual system will be, select the operations which are least likely to be fundamentally disturbed by future developments, and start.
2. Accept the fact that it cannot all be entirely foreseen, and that some of the revisions will be changes to revisions. We learn by doing, but preferably not by going in directions too far off the mark. Press departments and other metalbending operations are favorite places to start because setup times in those areas have to be short to respond to any kind of level schedule.

However, one of the universal "principles" is to design changes for complete setup flexibility of the machine. If it is to run 10 different parts, A through J, the best setup condition is the ability to switch quickly from any one of the 10 parts to running any other one. A modification that only provides a very short setup in switching from part A to part B may not accomplish much. By eliminating the variance in setup times between any pair of the 10 parts, we eliminate the tendency in production to stage inventory ahead of the machine to organize setups in a sequence that minimizes the sum of all the setup times.

Another "principle" is to review the purpose for which each piece of equipment is used. If a plant has five 200-ton presses, all of the same model by the same manufacturer, it is tempting to shortcut by making the same basic modifications to all five. However, one may be used for a large variety of stampings, another as a blanking press feeding another press, one for deep draw, and so on. They all have different objectives in use, so the details of setup required may be different. Setup modifications should be the result of a manufacturing process study, not only an equipment study.

2. Study the existing setup procedures

The regular study of setup methods becomes a standard activity in stockless production. As is true in many industrial engineering reviews, it is very easy to think a setup method is understood but overlook a small but vital activity. Therefore the current setup methods need detailed review from the first signal a setup is needed until every adjustment is complete and every tool returned to its place.

If no one has paid attention to setup methods before, a typical result of a setup study is that no two people setup the same machine for the same job in exactly the same way—and the results are not always equally satisfactory. In fact, if machines are set up infrequently, tool or machine damage sometimes occurs during setup. One of the fears is that if setups occur often, the cost of breakage will rise. Actually, just the opposite seems to occur. If people have to setup frequently, they learn how to do it properly.

A methods study of setup time follows the same principles as an industrial engineering methods study for anything else. The "correct" current setup procedure has to be established from among the variety of methods that may be in use. Then it should be analyzed to break it into short, distinct elements of work, and documented. This is organized into a format by which the purpose of each element is then determined, and ideas are proposed for eliminating or shortening each element. Exhibit 5–3 shows a format for doing this which is used within Sumitomo Electric Company. Some of the most obvious and easy improvements will be made almost instantly, but for planning and tracking the more difficult ones, Pareto analysis of the general type shown in Exhibit 5–4 may be helpful.

In a crash program to reduce setup times, not all the people making setup analyses may have a good background in industrial engineering methods studies. It is a good idea to give a format to work by and some basic instruction in methods studies, an enhanced version of the work simplification course. Everyone should work by an organized approach.

A common language among those doing setup analyses is also important. Pictures and diagrams are indispensable. However, both the setup descriptions and the analyses of them for time reduction are made much more clear if there is a classification system that makes distinctions between the following:

The machine itself. It can be described by any specific name, but that name should define it to mean *only* the basic equipment that remains unchanged during setups. That is not always as simple as it seems because sometimes a machine is changed so much that it is almost modified into something else during a setup, which in itself is the indication of a setup time problem.

Attachments. Dies, cutting heads, spray nozzles, clamps, hoses, adjusting arms—all of the devices which may be detached or altered in configuration during setup but which, when running, are attached to the machine.

Tools. These are devices used during setup for attaching, detaching, or adjusting; but when the machine is running they are not attached to the mechanism that directly performs the function of the machine. These can be further classified as common tools or special tools. Developing special tools is one way to cut setup times, but simplification can also lead to the reverse.

Material. That which is operated on to go into the product being made. A workpiece is the primary center of attention, but material may also be consumed. Welding wire is material, for instance.

The objective of the study is to compare the detailed objectives of the setup with the detailed study of the current setup methods on a particular piece of equipment used for a particular task. Then attack first in the easiest areas, those in which things can be done immediately to reduce setup time. Next, attack those areas which consume the longest time in setup, again doing the easiest things first. Save the challenging revisions until a number of simple changes have been made. Some of the simple changes may make the challenging ones less difficult.

3. Do as much external setup as possible—with the machine running

Work done while the machine is running is sometimes referred to as *external setup* or *outside setup*. Work done with the machine stopped is referred to in opposite terms: *internal setup*, or *inside setup*. The objective is to minimize the length of time required for internal setup. Experience is that switching as much setup activity as possible from internal setup to external setup reduces downtime by as much as 50 percent or more. (The terms *internal* and *external* are sometimes used with the reverse meaning from those given here.)

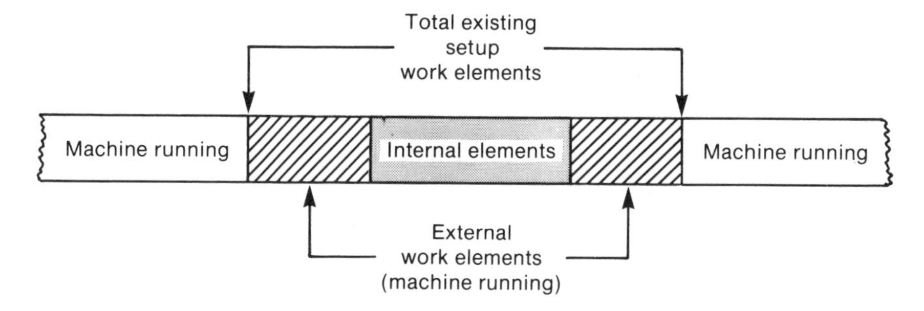

The core of a setup is the internal elements. The exchange of the attachments must generally be done with the machine stopped. All the attachments, tools, and workers required should be in place and standing by with everything laid out as if they were surgical instruments on a tray. With thought, workers soon learn to define the exchange elements in the narrowest way, so that every move made with the machine stopped is essential. As soon as the machine stops, the

EXHIBIT 5–3 Analysis sheet of setup work for face-grinding machine and examples of improvement

Operation	*Elemental operation*	*Tool*	*Time*	*External setup*	*Internal setup*	*Necessity of operation*	*Improvement idea*
Replacement of edger roll	Insert a rod in a hole on the side of the nut and loosen the nut by hammering the rod.	Rod, hammer			✓	To draw out the collar, To loosen the nut.	Isn't it possible to shorten the threaded part of the nut? (Idea 1)
	Draw the collar out of the roll shaft.				✓	To draw out the roll.	
	Lift the roll with both hands and draw it out.				✓	To insert the new roll.	
	Fit the new roll into the shaft.				✓	To reset it for the product to be processed next.	Fit or remove the roll without drawing out the nut? (Idea 2)
	Fit the collar on the roll shaft.				✓	To hold the roll.	
	Fit in the nut, insert the rod into the hole on the nut, and tighten it by hammering.	Rod, hammer	4 min., 10 sec.		✓	To fix the roll. To prevent loosening of the nut during operation.	

Matching of path line	Turn the feed screw by ratchet spanner and match the position of the pinch roll.	Ratchet spanner	2 min., 30 sec.		✓	Dimension of the succeeding product is different. To move it lightly.	Isn't it possible to move it quicker? (Idea 3)
	Turn the screw and set the position of the edger roll.	Spanner	7 min., 30 sec.		✓	To match it to the product dimension.	
	Loosen the nut, adjust the guide of the edger roll, and tighten the nut to fix it.	Spanner	1 min.		✓	To increase accuracy of grinding dimension.	Isn't it possible to eliminate tools? (Idea 4)

Source: Taken from page 9 of "Management by Standard Time and Work Improvement Through Saving of Set Up Time" (Paper circulated by Ryuji Fukuda, Sumitomo Electric Company, circa 1979–80).

internal setup should proceed without the slightest hesitation or confusion.

This requires organization of the external setup so that nothing is forgotten in preparation for the exchange process. It also stimulates the worker to mentally prepare for what he is about to do. However, standardizing the external set up work procedure on the same machine is also very helpful in causing workers to go through a set up by a routine which can be practiced and refined until there are no mistakes.

EXHIBIT 5–4 Analyze and act by the Pareto Principle

Example of Setup Change on Polymer Extrusion Die

Suppose that this type of setup in a company has never been formally studied before. It is complex enough and interacts with so many different considerations in production that it is hard to get started. So start by listing the major divisions of work in the setup, back-of-the-envelope style. Try to encompass the entire procedure, just as it has evolved to the present time, warts and all.

Estimate some times for each of these "megaelements" of work, even if the setup procedure is so ill-defined that this cannot be done very scientifically at first. Then divide the megaelements as best possible into times that are internal to the setup (machine stopped) and external to the setup (machine in production.) We might come out with some crude data that look something like the following table.

Time, minutes		*Which part of setup?*			
Range	*Average*	*Internal*	*External*	*Number*	*Megaelement description*
0–60	15		✓ (Always?)	1.	Locate all items for complete die assembly.
20–120	30		✓ (Always?)	2.	Assemble die assembly A for setup.
0–20	5		✓ (Always?)	3.	Check to be sure production run is complete.
5–15	10		✓ (Always?)	4.	Move assembly A from die department to machine.
5–15	10	✓		5.	Shut down extruder and rig for die exchange.
10–20	15	✓		6.	Remove die assembly B from extruder.
15–25	20	✓		7.	Mount die assembly A on extruder.
20–120	30	✓		8.	Preheat die A. Adjust induction heater.
8–12	10	✓		9.	Start up and lace material through draw-off equipment
10–120	30	✓		10.	Adjust die gap, draw speed, trim cutters, and so on.
5–15	10		✓	11.	Move die assembly to die department.
30–120	50		✓	12.	Disassemble, clean, and store die assembly B.
Totals	235	115	120		

EXHIBIT 5–4 (*continued*)

Anybody can refine this somewhat and begin to add detail, but in doing so the arguments begin. Bring out die A before removing B? Are we going to change with the heat on the extruder? Possible for some resins but not others? Die assembly A can go on any one of several extruders, each of which can run several products. Die adaptors are swapped among extruders and die assemblies. The heating coils do not always fit. The same tools and people are not always used. Some people want to set up a pet die. How many setups are occurring at once? Is the draw-off equipment being set up at the same time? Do we have advance warning?

Stop! If this continues, the analysis will disappear in a ball of fuzz. The objective is to analyze and act, not analyze and twaddle.

Most people looking at the above list can spot potential ways to cut setup time—ideas that will work most of the time, right now. We could build rolling rigs for *some* of the dies that would also allow the height to be preadjusted so that little height adjustment is required at the time of mounting. We might have standard heights some day, but not soon. We have needed to do something about the tangle of standard and special adapters anyway. We could make preheating the die a standard procedure, with some present exceptions. . . .

It will help direct the generation of ideas if we classify the megaelements by a Pareto analysis, as shown in the following diagram. A Pareto chart just classifies things by frequency of occurrence, largest item first, or, as in this case, by longest time required first. That draws the attention of the idea generators to the biggest blobs of fuzz first.

Take action. Whatever we know to do that will cut the setup times a little bit will also reduce the size of the problem a little bit, and that will help the further analysis of it. When we have accomplished several obvious or easy improvements, the setup procedure should be a little better defined; and now some of the megaelements can be further reduced to macroelement size, and we can repeat the same kind of analysis and action cycle on a more detailed level.

With a little more success behind us, we will even be able to define some of the macroelements into the sizes of elements that industrial engineers are accustomed to using for time study. The size of the problem has been reduced still more, and it is on the way to becoming standardized or repetitive enough that, in some instances, we can study a few minute details, or microelements.

Somewhere along the way, we reach a point where production problems other than a die setup begin to present more uncertainty than we can standardize into the die setup procedures. We may have to wait for the overcoming of some resin quality problems and so on, but not if we can help it. Why not go back and look at it a different way, say by redefining from scratch what is done to set up for particular products—different sizes and gauges of sheeting? The analysis and action cycle starts again.

Most of the methods for attacking setup time reduction are similar to those which have traditionally been used for attacking operating and transportation problems. Setup problems are only unique because they are sometimes so entangled with all the other problems of the production floor that we boggle in trying to clarify them.

What we want to do is remove "fuzz," occasionally called *variance,* from production. In this case, variance is not defined in exactly the same way as statistical variance, but it is related. Variance in production is irregular timing: uncertainty about what is being done or should be done; indecision about what to do next; the source of differences between plan and reality, or standard and actual; the interaction between more variables than we can get in mind at once. It is what happens when Murphy puts dirt in the end of the die adapter when the extruder die is set up. We would like to eliminate it. The process of analyzing and reducing setup times is a process of finding and eliminating a lot of it.

The exact format by which the Pareto principle is used to attack variance and setup times is unimportant. The determination to compress the ball of fuzz is important. During one of the author's presentations on this, one fellow grasped the idea perfectly by the remark, "Don't avoid setup; avoid screw-up."

EXHIBIT 5–4 (*concluded*)

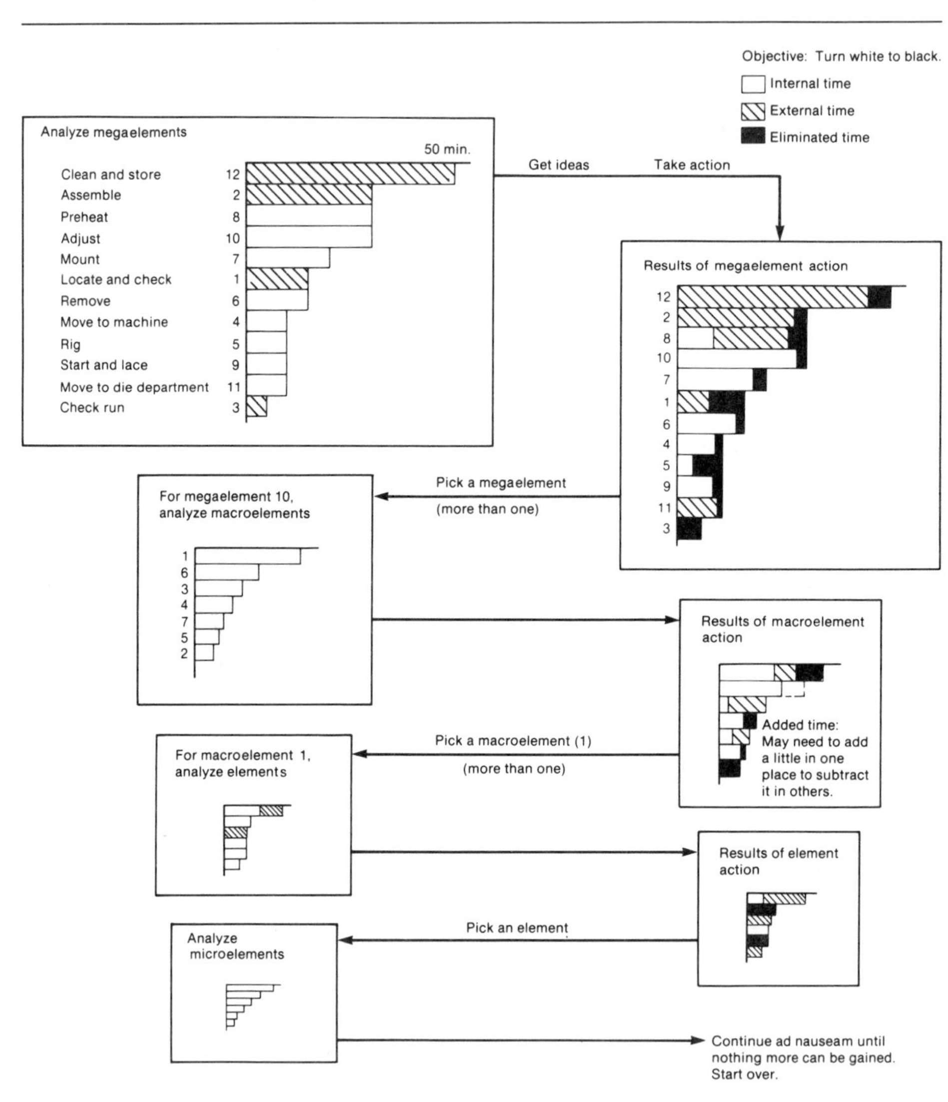

As many of the decision points as possible should be placed in the external setup, and as early as possible so there is little chance of disturbing internal setup. Decisions on whether cutting heads need sharpening should be made immediately after removal, not as part of the next preparation. Often this is a matter of communicating, among the parties concerned, such things as extrusion die clearances, tension settings, trial mixes, and the like. As setup experience develops, the need for making so many decisions is reduced because the procedure for setup becomes standardized. As much of the external setup as possible should be standardized, not just the internal setup.

It is not possible to execute a large number of rapid setups per day without modifying the equipment for it and revising the layout for the setups.

- Eliminate as much lifting as possible. For example, build tables with rollers to carry dies. Use pneumatic or hydraulic cylinders to push a die into position if it is too heavy to move manually. The objective here is to accomplish two things:
  —Eliminate the use of hoists or fork trucks. They take up space and consume rigging time.
  —Reduce effort required until the setup becomes a one-operator job. Then the operator can perform the setup without having to organize additional people to help. The more people required, the more difficulty in assembling a team for a setup at any time one is desired.

  After all this trouble to move heavy attachments on and off a machine easily, storing them on rollers is a good idea, too.
- For large equipment, presses with dies weighing tons, the same kind of thinking still applies, but it is not possible to make it a one-operator operation.
- Try to have enough space that major items of exchange can be moved quickly. It is ideal to move the new die into one side of a press and the old die out the other, but that cannot be done in all cases.
- If more than one person is required, develop a work pattern and provide the space.
- Revise equipment so that only small portions of large machine parts need to be changed during setup. Example: Attachments on the ends of push arms which move workpieces into position. Change only the attachments on the end, not the complete arms.
- Preheat attachments that are used with heat. Example: Dies for die casting. That eliminates trial shots for warm-up to working conditions as well as internal heating time.
- Do as much attachment and detachment as possible during external setup. Example: On an 800-ton press with a six-station die, the six-die sets were pre-positioned on a movable base. The exchange itself consisted of sliding an old base out and pushing a new one under the ram of the press.
- Avoid extensive threading or other time-consuming ways to mount and remove attachments:

—Cut off the excess length of bolts to be threaded.
—Cut away holes on one side so that only a few turns on the thread are sufficient to attach and detach.
—Use snap-over-center clamps.
—Use hydraulic or pneumatic clamps.
—Use quick-disconnect connectors for hoses, wires, cables.
—Get other ideas straight out of *Popular Mechanics*.

- Keep all the tools used in common on every setup in a location near the machine where each one can be found by instinct when needed.
- Keep all the tools and attachments specific to a given setup organized in a specific location as near the machine as possible to cut travel distance and to be sure that everything can be found when needed.
- Putting all the tools and attachments back in their assigned location immediately after a set up is complete is important for saving time and confusion during the next setup.
- If attachments must be exchanged with a tool room (for sharpening, cleaning, and so forth), this exchange should take place immediately after a setup so they are ready for use whenever next required.
- All tools and attachments should be cleaned after the setup and made ready for the next use.
- Keep the necessary workspace clear of obstructions whenever a setup is needed. To do this, establish consistent pathways for the tools and attachments separate from the pathways which cannot be quickly cleared of material.
- If material must be removed from access to the machine, provide for quick and easy access (movable sections of conveyor, collection hoppers on wheels, and the like).
- Keep gauges, reference patterns, and other items nearby so that checking the first piece from the machine requires little transport time.
- Put diagrams next to machines, or put labels next to parts of the machine or attachments, if useful. Put guide marks or lines where useful.
- Use color coding to quickly sort out hose connections or wire connections. Devise multiple connection plugs where useful.

In some cases the easiest way to achieve fast setups is for workers and material to move from one location to another. Example: Moving from one welding jig to another, which requires some adjustment in the method of workplace organization. Each work center and each workplace must be organized to accomplish fast setups by methods that may require some ingenuity and originality. The points given above are a guideline for typical cases. If setups are done several times daily, most of it boils down to engineering the setup methods to be repetitive operations, just as is true of running operations.

Once the importance of setup time reduction is accepted, things initially thought to be very difficult may become less difficult. The sketches in Exhibit 5–5 through 5–14 illustrate nothing complex and are intended only to start the idea process. Several U.S. tool manufacturers have over the years offered equipment featuring quick setup times, but the feature was not always popular, probably because setup time reduction requires consideration of the application of the equipment, layout, and shop floor methods, not just equipment design modification.

Most of the die-handling methods avoid the use of cranes and hoists, especially for positioning the die. It is slow positioning a die both horizontally and vertically when it is suspended, so the objective of most modifications is to maintain positive control of the die position during the exchange, and better control is obtained if the die comes to the machine at the correct height.

Layout modifications have fewer options for equipment so heavy that its existing location is literally cast in concrete. However, the dies or other items planned for use during the upcoming schedule period should still be located where they are easily obtained. The exchange times for dies on even the heaviest equipment can be greatly reduced through making those modifications which are reasonable, and keeping an open mind about what is reasonable. Companies having a large number of dies cannot keep them all close to the machines, and they cannot all be stored without lifting. However, those used the most frequently should be the easiest to access. Positioning the dies and material so as to respond to the pull signals expected during a schedule period is part of the preparation for operations during that period.

Developing an assembly line for assembly of mixed models requires positioning all the different assembly tools for different models so they are easily accessed, but the major function is organizing the line layout or assembly station layout for presentation of the correct material in the correct sequence to the assembly operators.

4. Eliminate as much adjustment time as possible

Adjustment time is almost always a part of internal setup time, and sometimes a major part, so attention to it is vital. After displacing as much internal setup activity as possible to external, experienced managers estimate that in the average case, about 50 percent of the residual internal setup time is trimmed by eliminating or reducing adjust-

EXHIBIT 5–5 Sketches and ideas for die exchanges in presses

Common method for light-to-medium presses

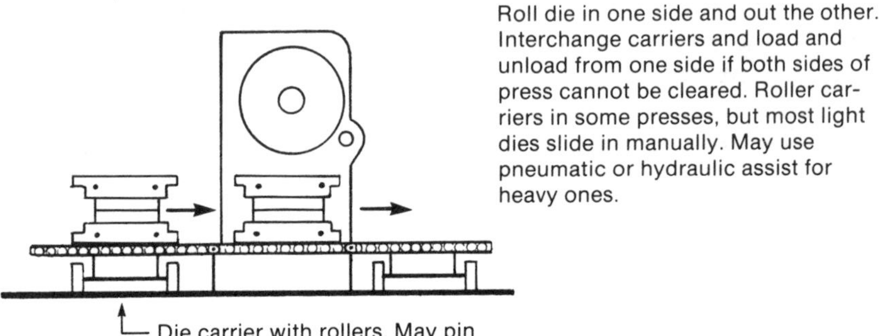

Roll die in one side and out the other. Interchange carriers and load and unload from one side if both sides of press cannot be cleared. Roller carriers in some presses, but most light dies slide in manually. May use pneumatic or hydraulic assist for heavy ones.

Die carrier with rollers. May pin to press. Suspended, swinging dies take too much time to control.

Store dies frequently used during production period on roller racks the same height as die carriers. (Rollers are too expensive for heaviest dies.)

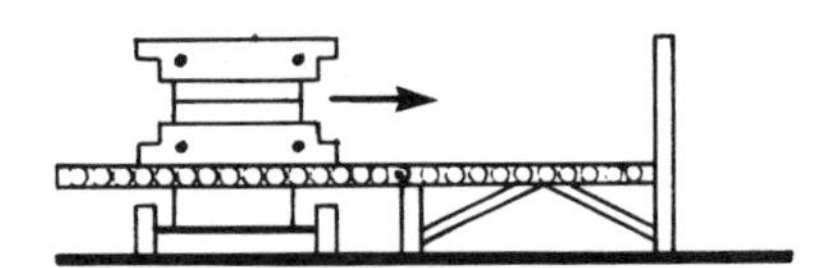

One idea seen for large press with multistation dies

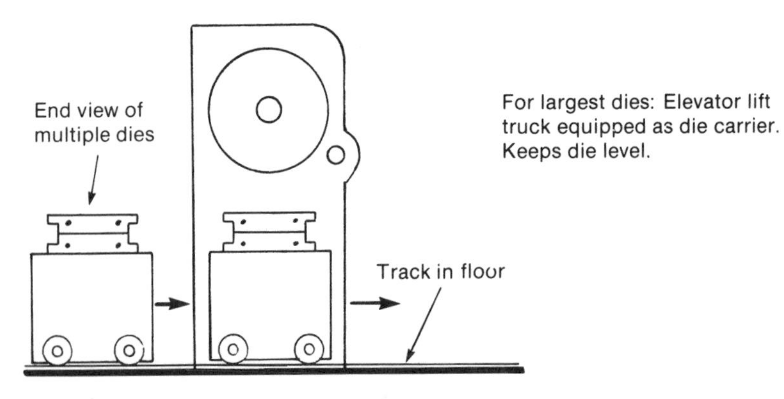

For largest dies: Elevator lift truck equipped as die carrier. Keeps die level.

Entire base of press, mounted with multiple dies, rolls under press on track. Can move less weight if only a common bolster plate is exchanged.

EXHIBIT 5–6 Sketch of quick exchange of polymer molding dies

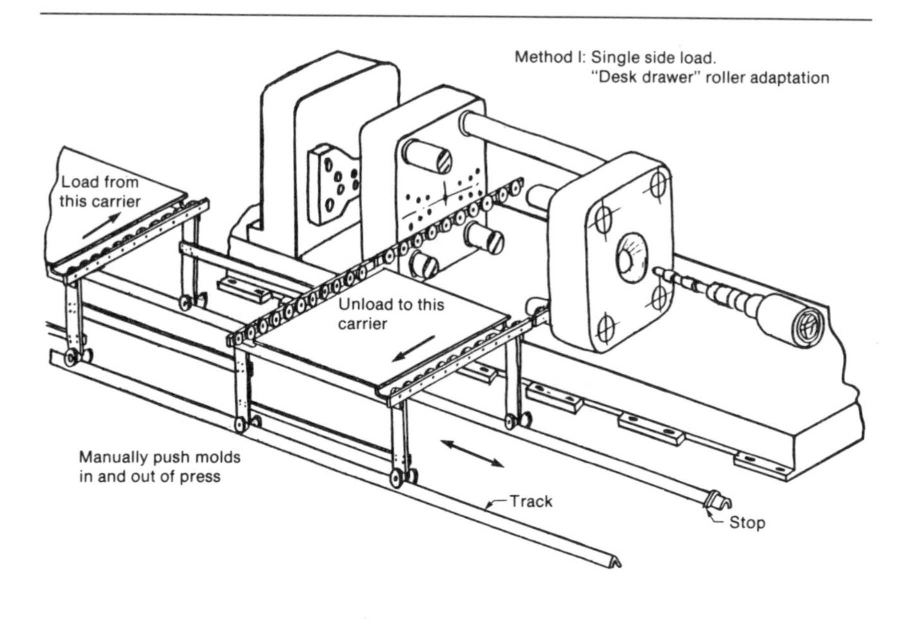

EXHIBIT 5–7 Sketch of quick exchange of polymer molding dies

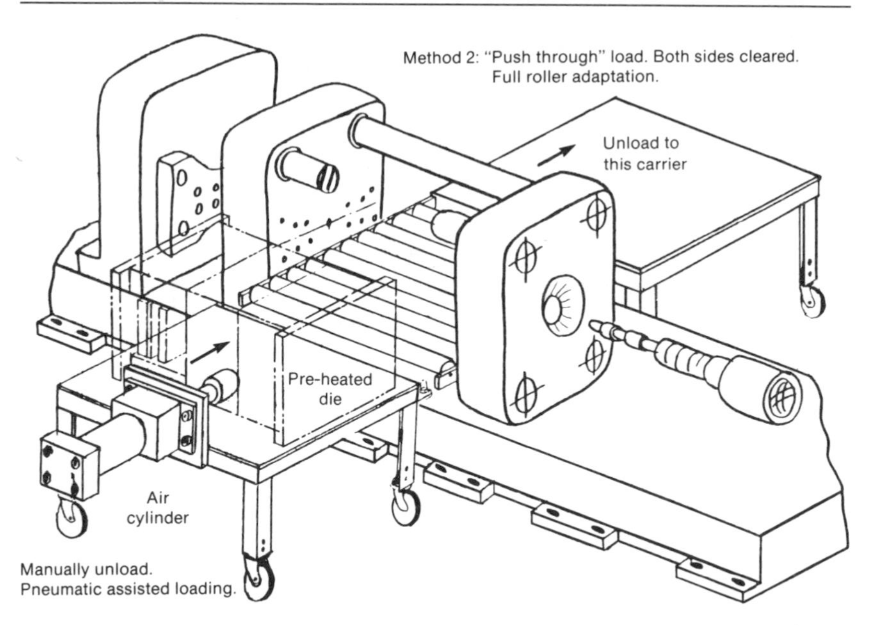

Many methods are possible. Methods are devised based on die weight and layout of area surrounding machine. (Illustration adapted from material provided by courtesy of General Motors Corporation.)

EXHIBIT 5–8 Die-positioning methods

Die positioning using locator pins

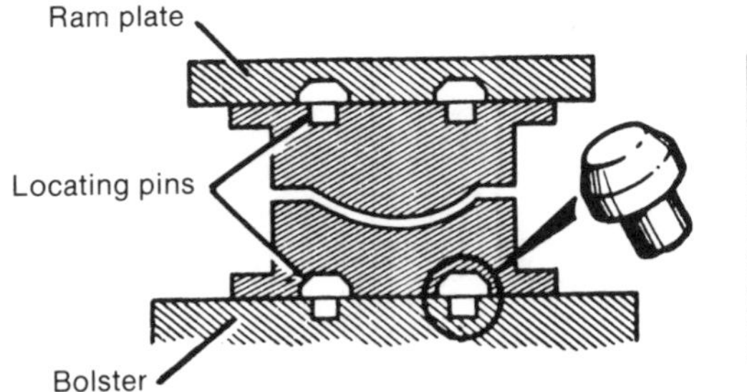

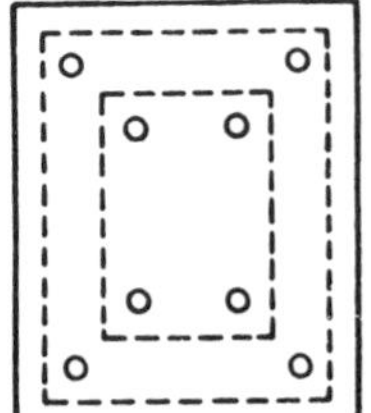

Hole pattern in ram plate and bolster. Common pattern for all dies and presses used in common.

Method using roller base and stopper. Many variations of this are in use.

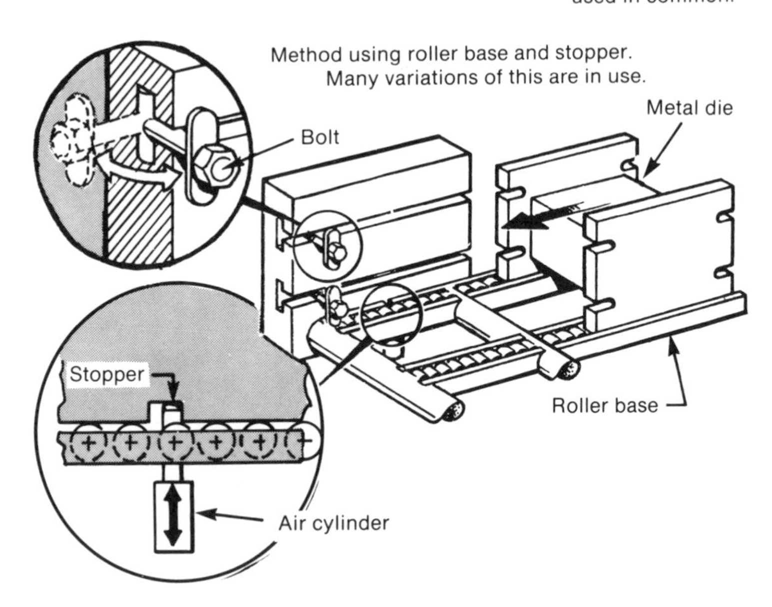

Illustration developed from material provided courtesy of The Bendix Corporation.

ment time. By eliminating the adjustment activity, we eliminate any scrap that it requires as well as the time required to dial in.

Some adjustment elimination is as simple as marking the exact spot on the floor to locate reels when feeding different types of coil stock to a press, but many adjustments require more technical consideration. The more a setup requires resetting a combination of temperatures,

EXHIBIT 5–9 Adapters for making press dies the same height

Adapters for different die heights

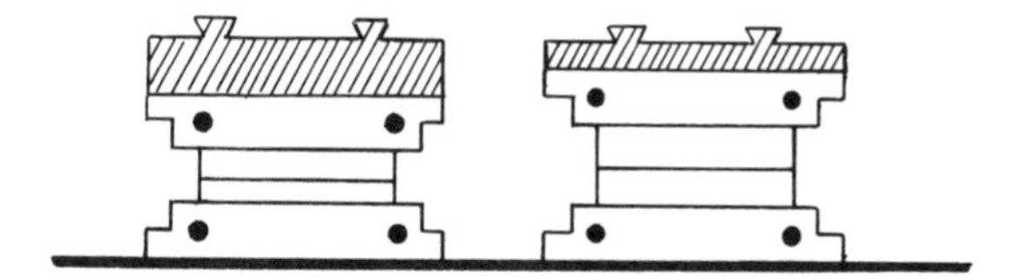

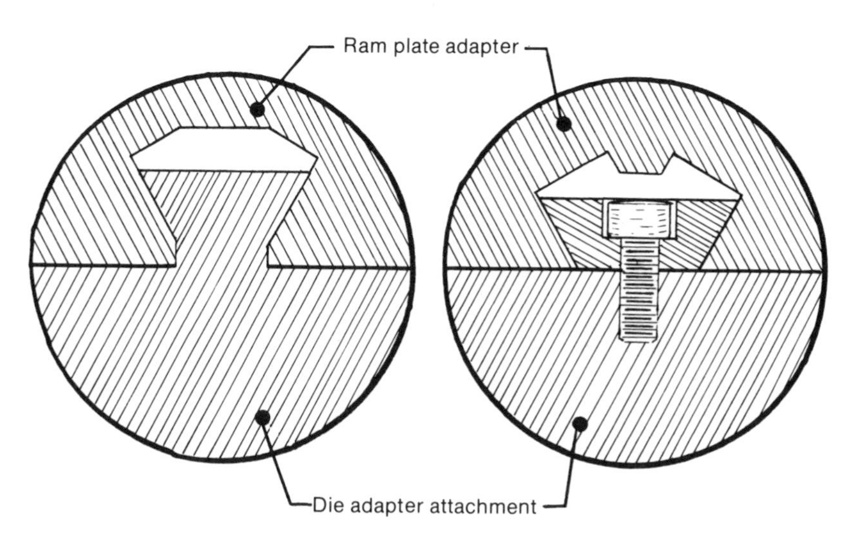

Adapters are important in cutting setup adjustments for many items, not just dies. V-slot adapters, sometimes used instead of T-slots, can allow quick attachment without great fabrication expense. Nonetheless, total expense may be too great if there are many dies. They may be considered if a few are used most of the time.

pressures, feeds, and speeds, the less simple the adjustment becomes, and process control can improve or deteriorate as a result of what is done.

The first step in this is to eliminate unnecessary positioning adjustments. The modifications of presses and die sets have this as one objective. Unless a company has specially designed dies, the external dimensions of different die sets are different, and they have different shut heights. The modifications to position a die set accurately in the X-Y plane normally consists of installing a set of locator pins or stops

EXHIBIT 5–10 Sketch of exchange of part-removal mechanisms

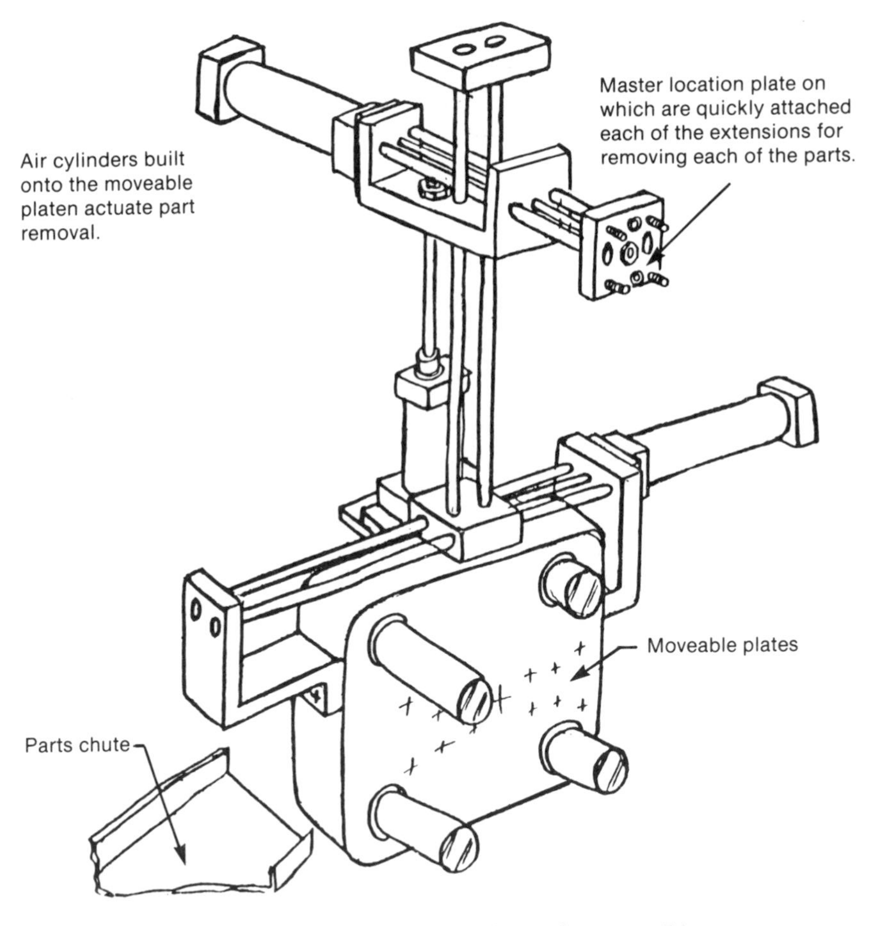

The principle is to exchange the smallest, lightest pieces possible.

which have the same spacing for dies of different sizes. Well done, a single push of each die set against the stops puts it exactly in position under the ram of a press.

If possible, die sets and ram plates can be modified with adapters so that the shut heights are equal for all, thus eliminating the time for adjusting the ram extension and also providing the means for quick

EXHIBIT 5–11 Sketch of master plate modification for quick exchange of multiple-head drill holder

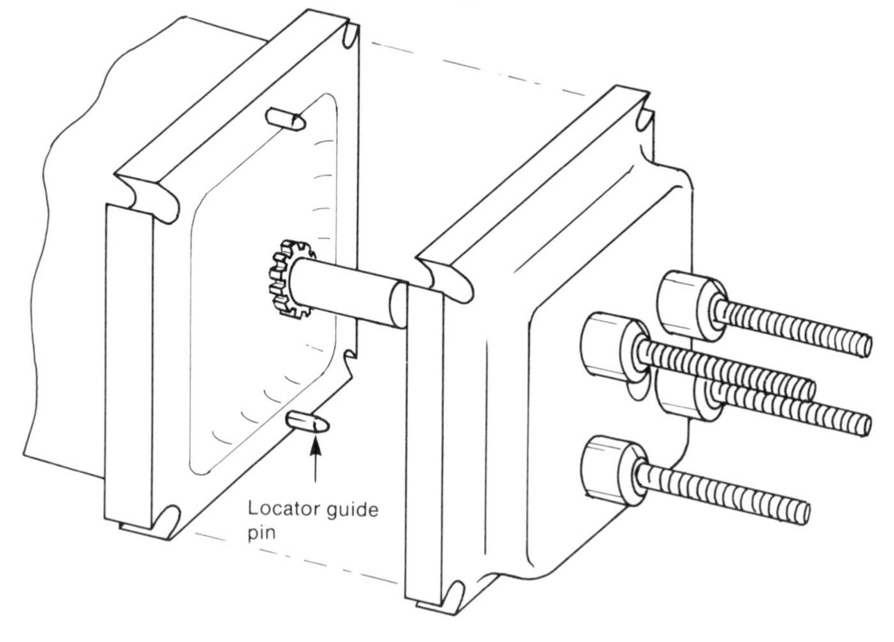

This type of concept for a master plate for quick attachment/detachment is sometimes used in the United States. The principle is to do as much of the detailed setup as possible with the machine running.

attachment of the die to the ram. A few ideas for die positioning and height adjustment are sketched in Exhibits 5–8 and 5–9.

If modifications to equalize shut heights are not physically possible or prohibitively expensive, the ram extension has to be adjusted. (Consider adapting at least the most frequently used dies to the same shut heights.) The time to adjust ram extension can be reduced by establishing a series of step-function stops on the drive which does this. That will at least eliminate zeroing in on height adjustment.

If die engineering or construction is poor, or if the dies are worn, the dies themselves must be brought into the tolerance ranges neces-

EXHIBIT 5–12 Sketch of layout for light presses located in department they feed

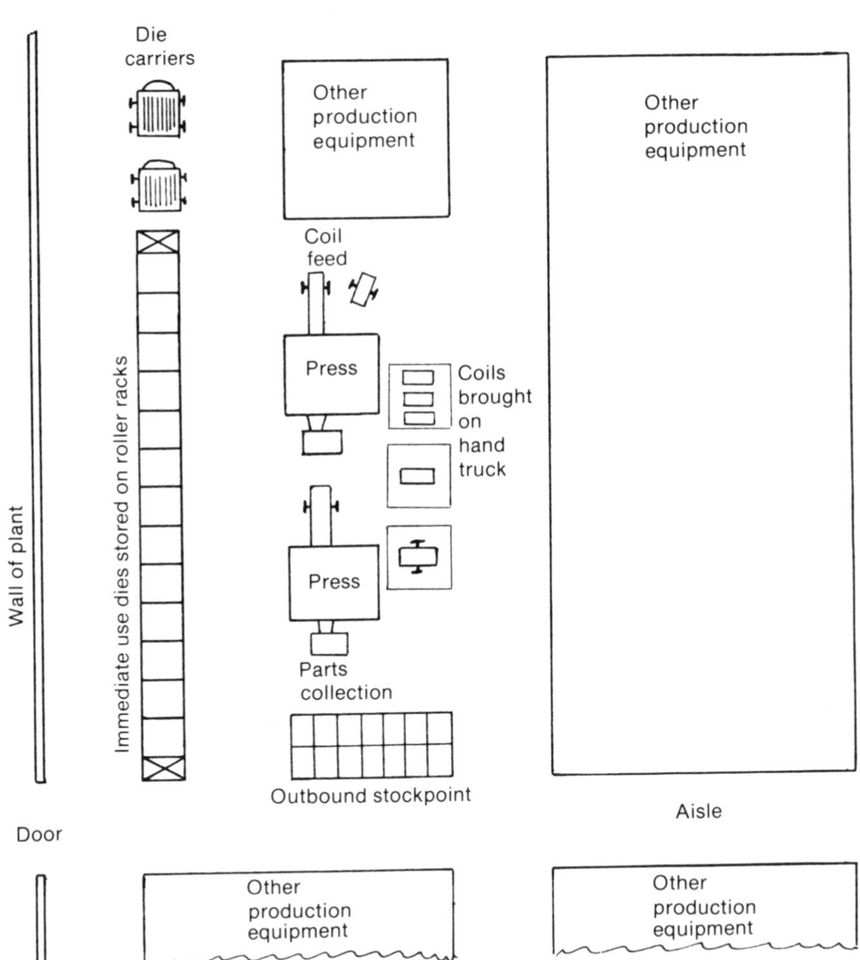

This is very near the ideal layout, but it depends on the presses being light and the noise level acceptable. If coils are light enough, they can be moved by handtruck and hoisted to the unwind stand by a garage type lift. If there are too many dies for them all to be stored at press level, devise a lift for the carriers.

sary to eliminate adjustment. If the die set can only be made to work by shimming and tweaking, a fast setup is not possible, so stockless production demands attention to the basics of manufacturing engineering and maintenance.

The results of most die setups and many other production processes

EXHIBIT 5–13 Layout of heavy press department

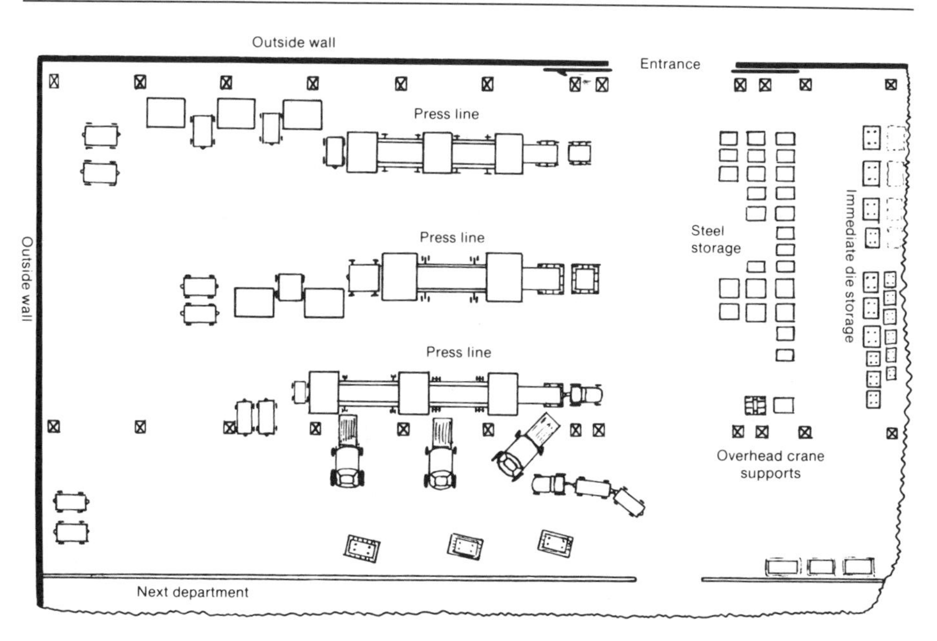

This is a much less ideal situation. The space restrictions and the positioning problems for dies and material during a multipress die exchange require careful thought. Over time, some of the obvious problems can be solved, but much of the initial effort is modification of dies and presses for quick exchange, plus putting as much of the collection equipment as possible on wheels and planning the location for it to quickly clear a path for the die exchange.

can be examined by keeping the last piece run on the same die and comparing it with the first piece run on the new setup. If tolerances are close or the piece complex, the first piece may need to be checked with check fixtures or gauges. If so, keep the fixtures or gauges close at hand so there is no long travel just to check a piece. Sometimes the check fixture can be stored with the die set, but they may be too delicate. Keeping them mounted on a wallboard near the press is one way to keep them at hand without taking much space. The objective is to eliminate adjustment so that the first piece made on a setup is good for production.

Proper instrumentation for temperatures, pressures, and speeds are important. Instrumentation can be expensive, so keep in mind the purpose of it—to be able to reproduce the setup. Instrumentation required to maintain process control may be more accurate and expensive, but that is a slightly different issue. It does not take complex

EXHIBIT 5–14 Layout of die casting department

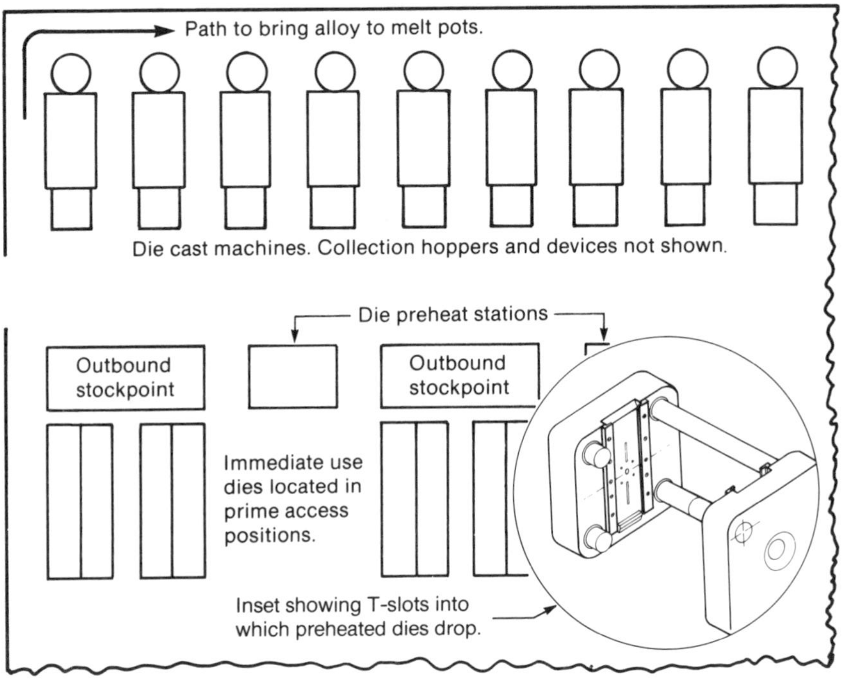

Path to bring alloy to melt pots.

Die cast machines. Collection hoppers and devices not shown.

Die preheat stations

Outbound stockpoint

Outbound stockpoint

Immediate use dies located in prime access positions.

Inset showing T-slots into which preheated dies drop.

Although suspending a die from a cable is not the best control, the heat in die-casting mandates the method. The die exchange is performed with a double hoist. Dies slide down the T-slots and into position against a stop. All dies for the same machine have adapters to position correctly against the same step. Dies are kept stacked in marked locations across the aisle from the die-cast machines.

measurement, for example, to set the pressure on a pinch roll that sets the tension on a web entering a process. It is only necessary to mark the adjusting knob or screw so that it can be quickly repositioned for the tension which has been found to work for different thicknesses. The little pieces of tape on the dials of various pieces of equipment are often an operator's attempt to mark where to reproduce setup conditions.

However, many pieces of production equipment are inadequately instrumented for setup. Gauges are missing or broken. Sensors are broken or inaccurate. Sometimes even the means of adjusting are crude, so one finds operators adjusting the depth of a milling process using pliers on the stem where an adjusting wheel has broken off.

The study of a setup should evaluate what really has to be measured in order to establish and reproduce the conditions of setup: speeds, temperatures, pressures, tensions, gaps, and the like. Instrument with sufficient accuracy to make and reproduce setups. The simpler the conditions of a setup can be reproduced, the better, but this is sometimes not quite so simple. If a setup has to compensate for variations in material thickness or density, it involves more technical issues of process control. A study to reduce setup times may uncover a number of these issues, raising anew, for example, the issue of proper temperature and pressure conditions for polymer molding. It directs attention to what is known and practiced on the production floor as opposed to what is known in a laboratory setting.

Many of the changes that will permit rapid adjustment consist only of marking setup positions so that they can be quickly reproduced without a search process. However, modification of traversing mechanisms for step-function adjustment also helps. (This type of adjustment is well known as the push-button method.)

The ways in which setup can be made easily and quickly are not confined to the ideas that have been used so far, and certainly they are not confined to those presented in this chapter, which is intended only to focus interest on the subject. Many companies that have never had a program to reduce setup times in production make in their own factories consumer products which have been designed to allow consumers to setup and use them quickly and easily.

5. Practice and refine the setup procedures

Executing a setup well requires correct reactions, so practice is necessary when the procedure is new and periodically thereafter when it is revised or when new people must execute them. Complex setups and those on large equipment must be done by a team, so some of this is team training. As a department begins to drop its setup times down into the minutes, the training and retraining for this becomes an integral part of everyone's training program.

In Japan, practice time may be set aside for the end of a shift after the schedule is met, or in special sessions at other times. The practice allows operators to build their skills and, equally important, to debug new setup procedures and try out new ideas. Setup practice for teams is a little like a football team working on a new play.

For team setups, practice should include assembling the team whenever needed. The team members may be scattered, especially if the setup comes unexpectedly. One reason for the level schedule is to permit a daily pattern to the setups so that they are ordinarily anticipated, but the system must respond to pull signals, not to a fixed schedule. In addition, the team members watch the signaling system of the lights. A warning light should signal when a setup is near.

As a plant moves more and more into stockless production and becomes more and more automated, the more an operator's daily routine includes a large number of setups. Constantly planning ahead to execute these setups well requires one to stay alert. If the running operations are automated, the last thing most operators should wish is to stand watching the machine run, but some operators may have difficulty accepting that their new role has more variety and thinking. Operator effectiveness seems better if people learn a moderately complex set of tasks to perfection and have input into their improvement.

What happens to setup specialists? Some equipment may continue to require a specialist, but, for the most part, who could be better at reworking equipment for quick change than setup persons? And once that program is well along, preventive maintenance becomes more important, so they can transfer there. Other jobs may also be open, but the point is that the objective of quick change is *not* to put the setup specialists out of a job.

Summary of setup-time reduction

Reducing setup time is a matter of concentrating ingenuity on the task over time. If we exercise patience in studying and replacing equipment, the amount of investment is not great. In fact, most companies did so much work that was absorbed into general departmental budgets that they have difficulty determining exactly how much many of the equipment modifications cost. The way to stifle the program is to create a separate capital budget request for every machine. By the time everything is reviewed and discussed, the program is slowed. If departments have only to stay within budget guidelines, they do not spend what the company does not have, and the machines where setup projects are useful are so obvious that no special study is needed. Just study the operations and do it.

Because of this, a company needs to have people capable of modifying equipment. Manufacturing engineers, tool and die and maintenance personnel should be able to perform most of the modifications. Turnkey modifications by equipment vendors may not turn out well because the changes required of specific machines are the results of the ongoing reviews of machine applications and production operations. It appears better to do as much as possible in-house, using the people who deal with the equipment and its idiosyncracies daily. Besides, if they do it, they take pride in it, rather than oppose it.

Some kinds of equipment do not yield very much to attempts to reduce internal setup times. Screw machines are difficult. Multiple-transfer machining equipment is difficult. In such cases, try to find less complex equipment to perform the same function. That cannot be done quickly, but companies with stockless production favor the selection of flexible equipment, and it is not unusual to replace several million dollars worth of "screw machines" with cells of simple machining equipment to do the same thing.

Such decisions are difficult, particularly if the equipment is not fully depreciated and is still operable. However, its cost is a sunken cost if more effective total operations are obtainable without it, and over time the usual result of stockless production is a decrease in the equipment asset base to support the same level of production.

LAYOUT AND HOUSEKEEPING

Organization of the workplace is essential if the plant floor is to become a replacement for the stockroom. Finding one part in one place is a material flow simplification easier said than done. Stockpoints need to be organized so that the location of each part is specified by number in a marked spot. This may seem like an unnecessary fetish, but as material begins to move more and more rapidly in small quantities, labeling locations is very useful even if workers have them memorized. It creates a stockroom mentality.

Likewise, if quick change is to become the norm, tools and attachments must also be organized. Making quick change a way of life is very much a matter of returning everything to its designated location after a setup is torn down. In fact, as many tools and attachments as possible should be put in their proper places *while* the setup is being torn down. A common way to do this is to place all tools and attachments on a skid or plate, which is then returned as a unit to its storage location. Before the put-away, everything should be cleaned or refurbished, ready for the next setup. The put-away needs practice and attention because it is really preparation for the next internal exchange.

In general, material flows laterally across a machine, so tools and attachments can flow from front to back. It may take several reorganizations before the pattern of flows becomes settled, and even then few of the flow patterns work out as well as desired.

Housekeeping is very much a part of this. Clutter is the enemy of fast tool flow and fast material flow. Dirt on floors and work surfaces is the enemy of quality production. How clean surfaces need to be is a function of the sensitivity of parts to foreign material, but all surfaces should be clean enough that there is zero damage from dirt, and no unnecessary delays for wiping and cleaning parts.

In most plants it is not necessary to be able to "eat off the floors." That degree of cleanliness is only useful when a clean room environment is needed, and employees may not accept a level of housekeeping that is beyond what can be shown to have an eventual purpose. Trivial harrassment serves no purpose, but employees should understand what is meant when rejects come from dirt and delays from the clutter. They often point that out themselves. Excellence in housekeeping with a purpose stimulates morale—the feeling that one is *expected* to do excellent work.

Layout and housekeeping are everyone's business. Every worker needs to understand how to organize his work area and why. Details of organization may be different for two different workers at the same work station, but if that is the case, they should discuss why this is true, unimportant individual differences or a source of inefficiency or process control problems.

The work rules given below come from the general rules for all production employees of Tokai Rika Company. The list is very similar to that given to employees of all stockless production plants. It directs everyone's attention to the importance of layout and organization of the workplace.

Stockless production work rules

Organize your area and your work. Classify tools, jigs, and materials to be used into three groups.

Things used frequently. Keep in the specified place ready for use at the required time.

Things used occasionally. Keep in the specified place and keep everything which is used for one operation together in the same location.

Things never used. Return these to the department from which they came or to stores. If useless, scrap them.

Put your work place in order.

Specify a location for everything.

Make someone responsible for each location.

Keep everything in order according to the general instruction on work place organization, and according to any specific work rules in addition.

Keep everything ready for use at any time.

Keep things clean. Things organized and put in order will be kept clean enough for immediate use at any time. Remove rubbish. Dust. Remove rust from the machines and oil them.

Adopt a positive attitude toward sweeping and mopping. Abandon the point of view: "*I* am the person who gets everyplace dirty, but *you* are a sweeper."

Maintain discipline. *Everyone* shall observe the rules on organization, putting in order, keeping clean, and sweeping.

Like most plants on stockless production, Tokai Rika has no janitors for the plant floor. Everyone is responsible for the organization and cleanliness of his work area. Stockless production requires detailed locations for everything, and it is impossible for the staff to prescribe or document them. Every worker must be able to perform this task, just as any good stockroom worker should be able to organize a portion of the stockroom. This is so essential that the top managers also observe the rules. The company's executive vice president says that he observes the rules and occasionally he makes it a point for employees to "catch" him cleaning his own office.

Housekeeping contributes to efficiency and productivity:

1. Fast and frequent setups.
2. Disciplined movement of material to keep control of the "depth" of the process.
3. A sense that quality work is expected on a quality product.
4. Continuous observance of the condition of tools, attachments, and machines; preventive maintenance.
5. The development of high-visibility methods of control. Strict use of standard containers, pull signals, and so forth.

Removal of inventory and clutter allows equipment to be positioned more compactly. This is desired to reduce transport distances, and it also frees space for other uses. Warehouse space is certainly available for other purposes, but space is usually made available on the factory floor as well. If most parts are put in small standard containers, the material handling is simplified. In many plants, fork trucks are unnecessary in the work areas themselves. Handtrucks are sufficient, and if the material can be transferred directly from machine to machine, not even those may be needed. The small containers can be moved manually.

A straight line layout all the way through production is ideal, but reality is that all the other limitations imposed by the current technology of production have to be considered. Dirt and noise from heavy fabrication have to be isolated from assembly (or vice versa); a part that requires painting generally cannot be painted where it is made, for example, so that most plants enter stockless production with a departmental layout much like any other plant.

Next best is streamlined flow through a fixed routing. Fixing the routings means that one part should be found in one place—come from a single outbound stockpoint. Therefore it should only be made in one place, with outbound stock points well-marked so that any employee can come find what is wanted without confusion. (Companies with many small parts attach a sample of each part to a circulating container. The material handler can compare a part in hand with the one on the container as well as refer to the part number on the container.)

The people in the work center supplying each part have the obligation to have parts ready when demanded, so if they have to scramble doing something unusual in another part of the plant to overcome their troubles, they are still obligated to place the part in the location where those looking for it expect to find it. That increases the pressure to perform as expected, not look for alternate routings.

As the work rules suggest, most of the difference in layout for stockless production fabrication departments comes within the departments and work centers, at least in the beginning. Only with purposeful development over time can operations be physically linked, as in the case history of the Kamigo engine plant, and have the result be what was really wanted.

The idea of one part being in one place and of everything else having its place is the opposite of randomization. The whole idea is to physically get material organized as parts flowing into final assembly as quickly as possible. Once this sequence of flow is established, do not allow it to be disturbed by randomizing parts again. Almost every part is made one at a time and used one at a time. Therefore the parts should flow from the point of fabrication to the point of use in as nonrandom a fashion as possible.

Because of this, small lot sizes do not add to material handling costs, provided material handling and material flow are revised to accomplish this. The cost of making more transport moves is more than compensated by making the moves shorter, quicker, and with less effort; and by eliminating double handling, jumbles, and resorts, beginning by not putting parts into stock and taking them out of stock. A good way to explain this is that each operation should come as close as possible to presenting one part at a time to be drawn toward the subsequent operations. In this way, all operations can think of themselves as a station in final assembly, even if they are remote from final assembly and not connected by a conveyor.

An increase in inventory between two work centers is a sign that this nonrandom material movement is beginning to collapse into confusion. Other signs are a decrease in tool organization, or an increase due to clutter or confusion, in the effort required to transport material. Variance is beginning to encroach on production.

In fact, if conveyors are used to transport material without exercising discipline over the limit of pipeline stock that can be in them, the result usually is that nonperformance somewhere causes the conveyor to be off-loaded and reloaded once in a while. Inventory in a conveyor is no more a sign of efficiency than if it were stacked on the floor or in a stockroom. Most plants going to stockless production have torn out their long conveyors, which had inhibited development of a disciplined limit on pipeline stock. Such control can often be better exercised through frequent movement of small containers.

The closer a fabrication operation is to assembly, the more important it is to feed assembly the mix of components in the sequence required, or at least to transport to assembly in the container sizes that are required. Many operations close to final assembly can be set up to feed in sequence even if space does not allow them to be a part of a final assembly line. Trim, paint, subassembly, and sort operations are typically of this type. If the pattern of setup changes in these work centers can track final assembly closely, the amount of material packed in the vicinity of final assembly can be greatly decreased.

The most flexible of operations can be directly connected to final assembly if space allows, but they must be able to set up in response to final assembly to do it. As an example, the author watched a simple brace being fabricated to order for different models on an assembly line. The brace needed a different bend and hole pattern for each model. The assembly operator pressed, drilled, and installed in about an eight-minute cycle. Only one part number was sent to the line. Several were installed.

Space at assembly does not permit that very often. The final part number has to be fabricated at a distance and shot to the assembly station, and this little example illustrates that point. The closer one gets to final assembly, the more the layout problem, in general, becomes one of managing a sorted flow of material through a confined space.

Assembly layout

To run a mixed sequence of models in final assembly, the assembly stations have to be designed with enough flexibility. Assembly tooling is generally simple enough that a setup change in assembly consists only of grasping the correct tool or switching to the correct assembly fixture. Tool planning does not usually require the space that fabrication attachments do, so usually there has to be some kind of ingenious rotational system to present different fixtures or different stations must be designated for different model fixtures.

Once the assembly station work layout has been resolved, the biggest problem is presenting an accurate, uninterrupted flow of material to the station. An assembly station needs to concentrate on assembling, not on material handling, and material handling usually means search and fumble. Material needs to be presented in small quantities as needed. The existence of cardboard boxes, unlabeled containers, and unlabeled locations is generally a sign that a great deal of parts search time is consuming what should be assembly time. The difference between good organization of small amounts of material and selection from a large amount of unorganized material in assembly can cause as much as a 50 percent decrease in assembly time, and sometimes the parts search time at assembly is not recognized.

An accurate presentation and positioning of parts at assembly stations is the key, whether the assembly stations are manned by humans or automatically. The pull system starts with the assembly station drawing what is needed, piece by piece, for installation.

The size of the unit assembled and the size of the parts for it are major factors. Large units (automobiles) are moved past stations at which only a few parts are attached. Small units (electronics) must dribble many small parts to each station. And there are all the cases and combinations in between.

To assemble a mixed sequence requires the ability to select parts in kits.[1] One of the simple ways to sort kits (of small parts) is from a color coded rack. Each pocket or bin on the rack is coded with one or more colors, so for the blue model, one takes the required number of parts from each of the blue bins. The same thing can be done electronically. A good design question is how much of this can be done at the assembly station and how much must be done as presorting.

For electronic assembly, the prevailing mode is to establish an assembly area feed station for assembly stations. Assembly stations can have kits fed to them upon demand by electronic signals or any other signals—the beginning of the pull system that will work its way back through the production network. The pattern by which parts are sorted and consumed at final assembly is a major factor in layout design of final assembly, and this in turn sets the pattern of parts consumption by final assembly which is transmitted to supplying areas.

Inspection, calibration, and "burn in" (of electronic gear) are all part of assembly. They consume space and sometimes material. Some of the kitting problems occur in these areas.

What we are working toward is the means to automate assembly. If parts attachment is to be done by robots or other automation, the automated system must be able to find the exact part required at the exact location programmed in order to work. The fabrication and supply system has to be able to feed that physically and organizationally.

There are a number of ideas for starting the layout of final assembly in this direction:

- Feed large parts in sequence by model. Gravity drops are sometimes an inexpensive, space-saving way to do this, as for a customized wheel mix to a truck assembly line.

[1] Packaging material and literature for various models are often the most difficult kits to manage in final assembly, and even several stockless production companies conceded not giving enough early attention to them.

- Smaller parts can often be presented in a sequence in their containers.
- Use inclined, gravity feed racks for standard containers. Material handlers feed the racks in sequence from the rear; assemblers select parts from the front.
- Use rotating turntables, or lazy Susans for feeding parts in sequence.
- Use transverse belts or feeder lines crossing the assembly line or parallel to it bringing material in sequence as demanded.
- Work stations can allow workpieces to be pulled off-line to a device that will allow them to be rotated or even turned upside down for a parts drop. This is possible, but takes ingenuity when running a cycle of mixed models with different dimensions.
- Use a gravity drop or auto-pick, which allows a combination of parts to drop onto a tray for easy selection. The worker does not have to reach into several bins or boxes while assembling.

Coordinated setup of equipment linked by material flows

During model change on an assembly line, equipment along the line must be set up at model change time. More correctly, for a flying change, the setups begin at the head of the line and flow down the line just ahead of the first unit for the model. If the line is assembling a mixed model sequence, the setup is almost indistinguishable from the rest of the assembly process; it is an integral part of it.

Extend the same idea to the setup of equipment for fabrication. As work centers and departments become more tightly linked, the more the setups at one location affect the setup routine at the work centers which feed it. The setups begin to occur in a coordinated pattern. As the fabrication becomes more closely tied to final assembly, the setups become synchronized to it. The layout of the plants needs to give some degree of consideration to development of the signaling system which allows this; the status boards, card racks, or signal stations and stockpoints are located so they have the requisite visibility.

As fabrication becomes so tightly linked that workpieces flow one at a time from machine to machine, the entire operation has to balance like an assembly line, and the objective of setup time reduction is

$$\text{Setup times for each machine in the series} \leq \text{Cycle time}$$

If this can be done, the setup causes no disturbance in line balance. The setup flows through the sequence of equipment just as if it were another workpiece. The only delay is the cycle time for one unit.

On an assembly line, in a group technology cell, or in any other series of equipment, if the model or the part being made is to change after every unit, then for all parts or models

$$\text{Setup time} + \text{Operation time} \leq \text{Cycle time}$$

Setup time and operation time become virtually indistinguishable, something possible in much assembly work. It is certainly possible if the only setup activity is a change of program in programmed equipment. For example, Nissan's Zama works is famous for robots in the body shop. The bodies are welded on an assembly line. At an early line station a scanner reads the body type, and this information tells each of the subsequent welding robots which programmed welding pattern to select.

In group technology cells, it may not be possible to drop the setup times to such short intervals as given by these criteria, but it is still worthwhile to operate the cells. These are the goals to which stockless production strives, and they may be possible either with very simple, nonprogrammed equipment or with programmed equipment.

6

Manufacturing process systems: Flow balancing

As inventory shrinks and operations become more tightly linked, production rates at all work centers need to become more closely coordinated. That is the purpose of reducing the inventory, and ideally, the cycle time of each fabricated part should be the cycle time of its use in final assembly. Balancing the material flow means also balancing the labor and machine times at all operations.

Most production people have some familiarity with line balancing. Traditionally, it involves mostly the balancing of labor at each line station. In the simplest case, supervisors and workers balance the work by shifting people and trading tasks along the line stations. For example, those doing custom-packaging frequently work with a different set of packaging every day. Each new line setup starts with a quick estimate of what elements of the packaging task can be done at each line station. If all operations are manual, it is easy to see where to make corrections for line balance. Watch where the material stacks up. The person at that point in the flow needs help, while others are underworked. Usually the workers can shift elements of work between them until a balance is reduced. This balance is not perfect, but they do not have a severe bottleneck while others have substantial idle time.

In the simplest cases, this begins to happen in parts fabrication as operations become tightly linked and are physically moved closer together. Parts are ultimately passed directly from one operation to another.

GROUP TECHNOLOGY AND BALANCING FABRICATION OPERATIONS

As a job shop converts to a more flow-oriented operation, one of the key steps is frequently called *group technology*. For specific long jobs, a group of dissimilar machines is set up in a cell so that all of them can be set up on the same job at once and so that there is little transport distance between them. The queues of work typical in a job shop are thus reduced, and overall efficiency is greatly improved. However, without the goal of stockless production, seldom is attention given to reducing the setup times of machines within the cell.

Many job shop supervisors and superintendents have used this method when work is overflowing. The major objective of expediting work is to move it through a job shop in exactly this way, setting up all the equipment on the routing so that it is ready just as the material being expedited reaches it. If the equipment is mobile, it may even be temporarily located together and dedicated to the job.

Group technology is widely practiced with stockless production. Three to 30 machines are located next to each other in a circular or U-shaped layout. Most of these U-lines consist of small, simple pieces of equipment such as drills, presses, grinders, spot welders, and so on. However, it is also done with other kinds of equipment. Tachikawa Spring Company set up sewing machines in circular layouts for sewing fabric for seat covers.

In a U-line for stockless production, operators move one workpiece at a time from machine to machine. They walk whatever path is necessary to coordinate activity so as to balance the line. Operators and supervisors work out a path for each person, and the result is a balanced load for each operator—except for one whose function in the cell may be designated for phasing out of existence.

The objective of operations in a U-line is to proceed as much toward full automation as possible. When machines are first assembled into the cell, the workers may need to position the workpieces in each machine, but as soon as possible, the machines are modified so that the workpiece positions itself with little or no adjustment from the worker. Sometimes the machine may be modified so that the work cycle starts itself as soon as the workpiece has settled into position, but usually the worker must hit a switch to start the work cycle. When the work cycle is complete, the machine stops itself and ejects the workpiece into position to be easily grasped for transfer to the next machine or to the outbound stockpoint. The cycle rate of the cell depends upon how fast the workers complete their tours around the U-line and return to the same machine to feed a new workpiece.

The diagram in Exhibit 6–1 shows two workers moving in two cleanly divided paths. However, anywhere from one to a half dozen

EXHIBIT 6–1 U-line layout

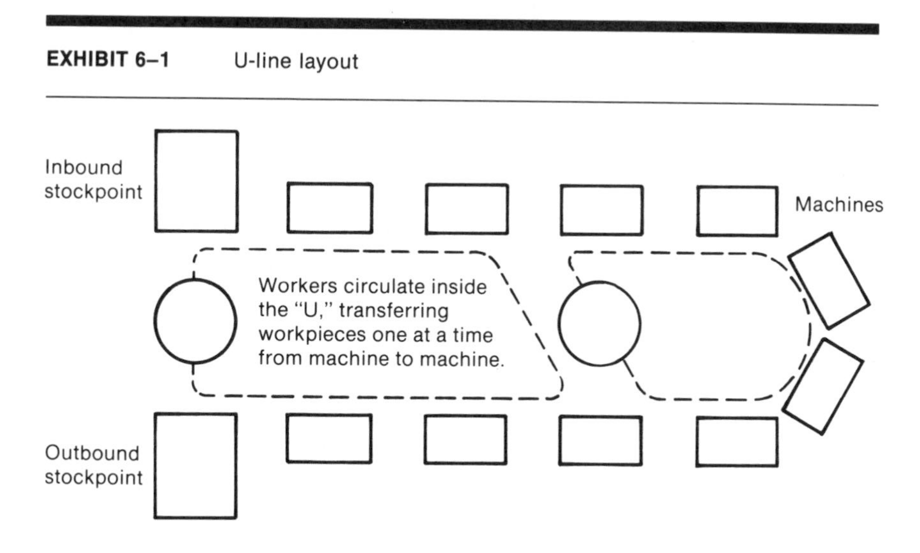

workers may move through separate paths, and sometimes the paths are not so cleanly divided, depending upon how they need to share their duties in order to balance the workload. If too many people work inside one of these U-lines, it becomes difficult to evolve a balanced work pattern and also difficult for them to avoid bumping into each other.

The output of the U-line as a whole needs to be balanced to the rate at which the parts it fabricates are consumed by subsequent operations, and eventually by final assembly. Separate parts fabricated by the same U-line may not require all the machines set up in the line, and they usually require different amounts of total work time, but all the parts made in one U-line should require labor times in the same general range so that the U-line does not require a different number of workers for each different part or difficult rebalancing of tasks for each part.

The overall output rate of a U-line can be adjusted by adding or subtracting the number of workers in it. However, during a given fixed schedule period, one would like not to have a complex series of manning changes. When the fixed schedule period changes, manning patterns may change, and sometimes there will be a revision of the number of machines included in the U-line. This is necessary to reestablish a work pattern in the U-line that will match the cycle times of the parts coming from the U-line with the cycle times that will be required by the new schedule.

People who have experienced difficult line-balance problems with equipment fixed in place and with narrowly trained operators may find this degree of constant rebalancing difficult to accept. However, the U-line is established with simple equipment modified for flexibility and with workers cross-trained with multiple skills. A skilled operator can learn to operate and set up every machine in the U-line.

The supervisor is primarily responsible for maintaining such data as are needed to evaluate how many people are required for different cycle rates in a U-line. This is not a task requiring a huge amount of data, but a set of rules based on experience. It should be compared to a situation in which a supervisor rebalances an assembly line by trial and error and a few calculations. Similar on-the-spot rebalancing happens in production every day for assembly lines having 30–40 people or more, provided the nature of the work allows flexibility to move people and shift tasks. Naturally, the supervisor can get help from professional industrial engineers when confronted with complex problems.

The balancing of a U-line is easier when the cycle time is over 30 seconds. Typical cycle times are 30–180 seconds. If the cycle times required are less than 30 seconds, start a second U-line. This requires twice as much equipment, but the equipment for U-lines should be kept simple and therefore inexpensive. In addition, if the volume grows to a level at which cycle times have been reduced to something under 30 seconds, the equipment may be justifiable.

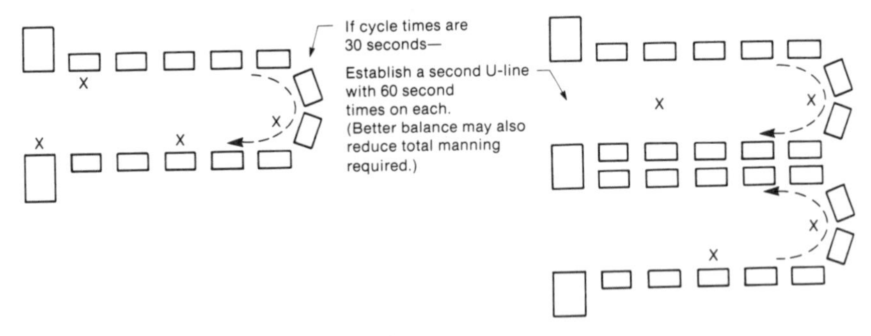

A very important principle of stockless production shows up in this. *Workers are the most important resource, and they should be the limiting factor.* The equipment is only going to be used to produce the amount of product required anyway. Using it to produce more than this is a self-deception because eventually inventory will build up somewhere in the production system, and the equipment must be stopped. In addition, the objective is to also reduce the amount of labor, and this is done by balancing operations. How the equipment is used is what is critical.

The shift from a single U-line to two U-lines may also result in reducing the number of workers because balancing is easier. To do this requires concentrating on removing elements of work from one worker until he or she is no longer needed for that task and can be given a new one. The same principle can be applied even if the U-lines are not divided.

Exhibit 6–2 illustrates the general case for attacking productivity through line balancing in a U-line or in any other configuration of people working together in a tandem arrangement. The objective of productivity improvement is not to reduce the cycle time if that is unnecessary for meeting the schedule; it is to reduce the amount of labor required while producing at a cycle time that produces parts only at the rate needed. Productivity improvement which is only stored inventory must be compromised at some point because the machines must stop while subsequent processes work off the inventory.

Depending on the tasks, group technology in the form of U-lines comes in a variety of forms. The essential features are the grouping of simple, flexible machines to eliminate inventory, floor space, transport distance, and quality problems. A U-line very often provides superior performance to multistation, high-speed automatic equipment because the U-lines can be set up more quickly, thus promoting production flexibility. They are not so hard to develop and debug.

Set up in accordance with stockless production, U-lines or group technology cells have several major advantages:

- Inventory between the operations is eliminated.
- Visibility is promoted. An entire sequence of operations is compactly located and can be studied as a whole.
- Quality is promoted:
  —One or more stations in the cell can be for inspection if need be so there is no transport and delay for inspection.
  —Better yet, equipment in U-lines (and elsewhere) can be set up with check jigs or measuring instruments to test for completeness or correctness of operations on the same machine primarily intended to perform a conversion operation. If checking and measurement can be built directly into the fixtures which hold a part, it can be inspected either just after the operation being inspected is performed, or as a part of the subsequent operation. This can be done if it does not make machine setup too complex to be quickly done.
  —Even without specific inspection by measurement, immediate feedback and informal checking are built into the direct trans-

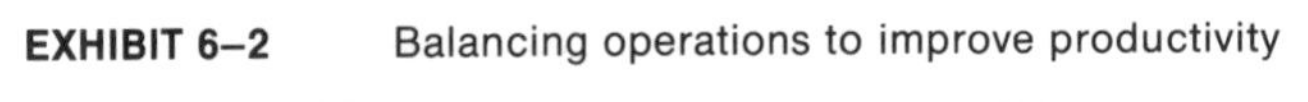

EXHIBIT 6–2 Balancing operations to improve productivity

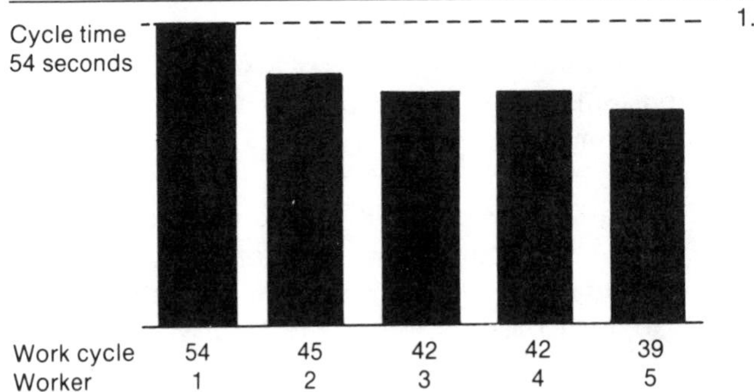

1. Original situation

   Fifty-four-second cycle time is adequate.

   Five workers.

   Worker 1 shows major work imbalance.

   Paid time per cycle: 54 × 5 = 270

   Idle time = 48 seconds per cycle.

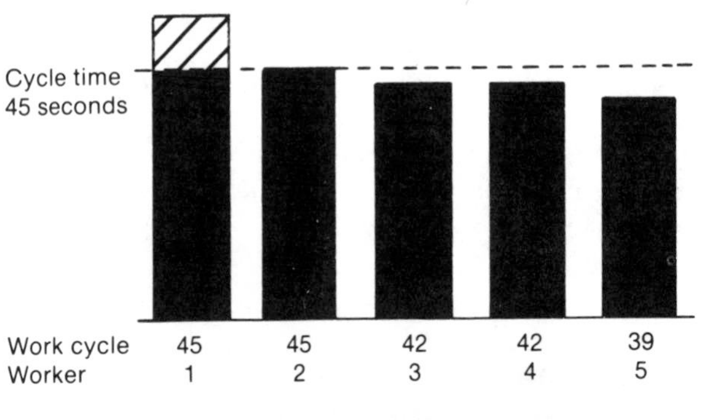

2. Typical approach

   Attack bottleneck. Cut nine seconds from Worker 1's time.

   Cycle time: 45 seconds.

   Presumed benefit: Paid time per cycle drops to 45 × 5 = 225 (17 percent decrease).

   Idle time down to 12 seconds per cycle.

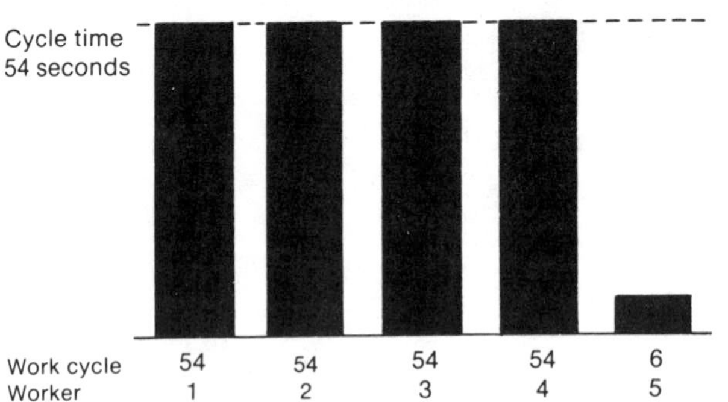

3. Stockless production approach

   Balance four workers to 54-second cycle.

   Then attack the remaining six seconds of time for Worker 5 to eliminate need for worker 5.

   Cutting six seconds of time will yield a paid time per cycle of 54 × 4 = 216 (20 percent decrease).

   Since there is no *need* to reduce cycle time, the 17 percent reduction by the typical approach only builds inventory somewhere.

Adapted from an illustration by Jidosha Kiki Company and The Bendix Corporation.

fer of parts from machine to machine. This promotes the self-detection of defectives by both operators and machines, and if the equipment is simple, the causes of defectives are not hard to find and correct.

These features are characteristic of stockless production in general, and several more are illustrated by their prominence in the U-lines:

- Workers also stay more alert by rotating through a variety of tasks than when doing repetitious ones having a short cycle.
- Workers have more flexibility and reach when moving, so workplace organization can have a wider scope.
- One company, Tachikawa Spring, reports that workers became 12 percent more productive strictly on the basis of increased worker alertness when moving about, rather than sitting, in a group technology cell.
- The worker moves the material automatically as part of the task. No special material-handling equipment is necessary (although the development of mechanical workpiece transfer devices is desired if the U-line is to evolve into full automation.)

Exhibit 6–3 diagrams a specific U-line form of group technology at Jidosha Kiki Company. It shows the need for small, simple equipment. Note particularly the miniheat-treat device, a small induction heater. This prevents transporting the part to a heat-treat department

EXHIBIT 6–3 Group technology at JKC

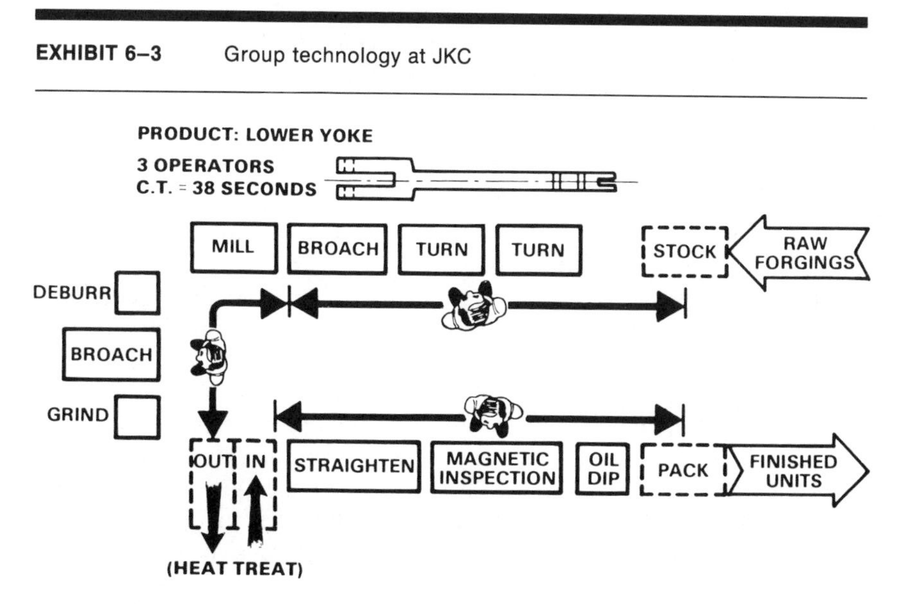

Courtesy of The Bendix Company and Jidosha Kiki Company, Ltd.

and back again. Heat treating is often regarded as a nemesis of streamlined parts flow, so some imagination is used to heat-treat parts in other ways.

Some U-lines may be dedicated entirely to the production of only one part, but most must be set up to run several different parts. Many of the setup changes do not involve radical changes in the type of part produced. For example, the lower yoke U-line in Exhibit 6–3 may produce several different sizes or variations of lower yokes. A 5–10 minute downtime for a setup change is typical of U-lines, and, Jidosha Kiki, for example, reports that the three operators of a 30-machine U-line take no more than 15 minutes for a setup, and that leaves much room for improvement.

If the setup times of all machines in the U-line can be reduced to less than cycle time, an operator can tour the machines, setting them up with little more delay than skipping one piece in the part flow. If setup time plus run time is less than cycle time, there is a possibility that the U-line could change setups with every tour of the line, but that depends on how many workers run the line and how the process balances out.

The eventual objective of U-line is to study how to directly link the simple equipment together so that workpieces are directly transferred from machine to machine, thus achieving full automation. However, this takes considerable study as to its worth. For example, it might take the attention of as many workers to keep a fully automated line running as would be true if it remained in the semiautomated state. The same kind of equipment may not be useful and it might lose its flexibility if fully automated. As full automation develops, it should come without these unpleasant problems if operations have matured toward it.

BALANCING THE PLANT

The U-line is the prime example of changing operations to balance them, but it is not the only example. All the fabrication areas linked together are subject to balancing. If electric motors are assembled with a cycle time of two minutes, then the casings for the electric motors should have a cycle time of two minutes. There are subtleties to this. Saying that the casings have a cycle time of two minutes is more demanding than saying that they are produced at an average rate of 30 per hour. That might mean that 30 are produced in one minute once each hour, or even that 240 are produced once in every eight hours of operating time. A cycle time of two minutes implies a uniform steady rate of production. That does not happen, of course, but by using the cycle times demanded by final assembly as the drum beat of fabrication, the emphasis is on how much fabrication deviates from this rhythm. Production improvements are based on marching everything closer to the same beat.

Traditional line balancing refers to balancing the labor at each station. In both traditional assembly line work and in stockless production, balancing the line or the plant also refers to balancing the material flows, but that is done by the pull system of material control. However, material balance and labor balance are not independent, because the primary sign of imbalanced labor is excess material or shortages of material. The amount of work in process between work centers is an indication of the imbalanced cycle rates between them, especially after the setup times have been reduced until lot sizes are small.

In the absence of stockless production concepts, balancing an assembly line usually has two goals:

1. Assuring that the work time possible at each line station is within the cycle time allowed by the line rate.
2. Minimizing the slack time at each work station by shifting work content between stations so that the work requirements at each are nearly the same and all are approximately equal to cycle time.

As we saw in the discussion of U-lines, this is *not* the objective of assembly line balancing in stockless production, but we prefer to set the line cycle time at the level required by the schedule and shift tasks so that they are taken away from operators who can then be moved completely from the line and assigned other work. This is also true when balancing entire plants.

In more complex cases of line balancing, models may be assembled in a mixed cycle as is true in auto assembly. Then line balance tries to provide a balanced average work time at each line station, so that the extra time or effort required for one model is compensated by the slack time permitted by the following model at the same station.

In the more advanced cases, line balancing must be precalculated by approximate methods in order to make an initial line setup which can then be refined by operational experience. These calculations may be done manually, but in complex cases they are done by computer. The usual objective of these in traditional line balancing is

$$\frac{\text{Sum of the work cycle times for all stations}}{\text{Cycle time} \times \text{Total number of line stations}} \leq 1.0$$

"Perfect" balance is reached when this ratio equals exactly 1.0, never really attained for any complex situation. The typical American method for balancing is to eliminate the work content at the bottleneck operations or shift it to nonbottleneck operations until the ratio is as near 1.0 as possible. Note that this is basically a method to improve labor efficiency.

In stockless production, the objective is to

1. Set the overall cycle time so that the rate of production is balanced with the rate of consumption of the parts being made.
2. Adjust the work content and work cycle times at each station until work times match the cycle time as nearly as possible.
3. Try to offload the work content of selected stations until they are no longer needed.

This approach may not look very different, but it is a major difference in the philosophy of production. Another helpful hint: When removing individual workers from an operation, remove the most skilled and flexible. They are the easiest to transfer to other jobs and the most likely to improve the tasks to which they are transferred.

As with almost everything else in stockless production, a formula may be a guideline, but only toward a careful study of detail. Balance is improved by reviewing exactly how to change operations to obtain it. That includes not only the cycle times, which can be run by different arrangements within work centers, but also the kind of material transport system and the total layout. The example in Exhibit 6–4 was concocted to illustrate some of the issues.

The example is not complex, but describing it becomes extensive anyway. If nothing else, this shows the pressure created by stockless production to examine production operations in detail. The work center is very tight in run time, so it has very little slack to respond to problems. If it becomes unsynchronized from final assembly, the extra setups required would throw it into overtime for the day. A breakdown lasting an hour or two would throw the workers into a mad scramble, and a quality problem would do the same.

The example may leave the impression that the work center operates according to a preset schedule, but actually it operates according to the material taken by the material handler. However, the work center cannot function in a mad scramble all the time, so what it also illustrates is that the entire plant from assembly line on back must normally function in a very coordinated pattern. Scrambling to overcome mistakes will throw off the material handling pickups, and one can see that if that happens, the signaling system will suffer a blip, and through that the problems of the work center will compound.

Suppose the setup times could be cut in half. If that were done, each part could be run six times daily rather than three, but in order to take advantage of that the transport times would have to occur more frequently. If a part is made six times, but only picked up five, material from the previous setup would still be in the outbound stockpoint when material from a second setup was complete. For those not yet into the thinking of stockless production, that might indicate that setup times have been reduced as far as it is profitable to do so. What it

EXHIBIT 6–4 Operations balancing: example of one stockless production work center

The problems of operations balancing are illustrated by considering a simple example of just one work center fabricating only three parts, each one of them a variation of the other. The plan for operating the machine at this work center might look like this:

Part number	*Number required per eight-hour shift*	*Setup minutes*	*Run-time minute per piece*	*Container size*	*Containers per shift*	*Cycle time of use**
P-1	500	5	.25	20	25	.86
P-2	400	5	.30	20	20	1.08
P-3	300	10	.40	20	15	1.43
Totals	1,200	20			60	

Time per shift available for work

480 minutes total
−30 minutes, lunch
−20 minutes, breaks
430 available minutes

Time required to run all parts on the average shift

P-1 500 × .25 = 125 minutes
P-2 400 × .30 = 120 minutes
P-3 300 × .40 = 120 minutes
365 minutes

Setup time available during the average shift:

430 available minutes
−365 run minutes
65 available setup minutes

Setup sequences per shift

65 minutes/20 minutes = 3 setups for each part

Plan for one setup sequence

P-1 9 containers = 45 minute-run
5 minute-setup
P-2 6 containers = 36 minute-run
5 minute-setup
P-3 5 containers = 40 minute-run
10 minute-setup
141 minutes, sequence

* The work center must respond to the demand for parts by subsequent work centers. These cycle times are based on the assumption that the parts are used in final assembly and that the final assembly line is running a repeating sequence of models using each of the three parts. Hence the models present a level demand for use of each part according to the frequency of its use on the final assembly line. The cycle time of part P-1 (430 minutes/500 parts) assume that final assembly has the same number of available minutes as the fabrication center during each shift.

EXHIBIT 6–4 *(concluded)*

Possible setup sequence for shift, with number of containers in each sequence

	1st	*2nd*	*3rd*
P-1	9	8	8
P-2	6	7	7
P-3	5	5	5

If the demand for each part in final assembly is exactly as planned, the number of containers required for each part is not much above that required to hold the parts for the inventory buildup which occurs when each part is run. Consider, for example, part P-1. If eight containers are filled in the first run of the day, then at the time of starting, only one container needs to be at final assembly. At final assembly one container lasts

$$20 \times .86 = 17.2 \text{ minutes.}$$

At the work center, the first container will be ready:

$$5 \text{ minute-setup} + 20 \times .25 = 10 \text{ minutes after starting.}$$

So if the first container can be transported and made ready for use at the final assembly line in 7.2 minutes or less, only 9 + 1 = 10 containers are needed. This assumes that operators or material handlers will be on the spot at exactly the right time to move one container—something that may not happen, so one or more extra containers may be necessary at final assembly to cover the material-handling timing.

A very brief and simplistic simulation of the transport of containers interacting with the work center is shown below. It is based on material handlers taking material from the outbound stockpoints five times daily, every 86 minutes during the working part of a shift.

				Balance on hand outbound stockpoint (number of containers)			*Taken to final assembly (number of containers)*		
*Event**	*Setup in minutes*	*Run*	*Cumulative minutes*	*P-1*	*P-2*	*P-3*	*P-1*	*P-2*	*P-3*
Start			0	5	7	5			
(9) Finish P-1	5	45	50	14	7	5			
Pickup			86	9	3	2	5	4	3
(6) Finish P-2	5	36	91	9	9	2			
(5) Finish P-3	10	40	141	9	9	7			
Pickup			172	4	5	4	5	4	3
(8) Finish P-1	5	40	186	12	5	4			
(7) Finish P-2	5	42	233	12	12	4			
Pickup			258	7	8	1	5	4	3
(5) Finish P-3	10	40	283	7	8	6			
(8) Finish P-1	5	40	328	15	8	6			
Pickup			344	10	4	3	5	4	3
(7) Finish P-2	5	42	375	10	11	3			
(5) Finish P-3	10	40	425	10	11	8			
Pickup			430	5	7	5	5	4	3

* Numbers in parentheses show containers made exactly according to plan.

This situation does *not* represent any ideal way to transport or to establish a balance with final assembly. It does not show the work center following pull signals from final assembly, but assumes a perfectly level use of all three parts at final assembly with no change in mix. This is only intended to illustrate some of the problems. To continue the illustration, see Exhibit 10–3.

really illustrates is that more frequent transport is needed. Never lose sight of the long-range goal.

To transport material more frequently without increasing material handling expense means attention to plant layout and organization of material handling circuits. Can the work center be moved closer to the assembly line? can the pattern of pickups be changed to permit more frequent pickup by other material handlers traveling a different circuit? Even if this cannot be done in the short run, there may be an advantage in leveling the production of parts throughout the day because of the beneficial effect on other work centers which feed into this one.

The effect of the material transport system on the system inside the plant is obvious from reviewing the minisimulation of Exhibit 6–4. Putting in a conveyor system to move material is one way to deliver a steady supply of parts, but conveyors also hold a great deal of inventory if they are very long. The real objective is to reduce the transport time to the assembly line. A robotic truck making rounds might accomplish that. So might a conveyor, provided it moves material quickly and does not have the capacity to accumulate vast amounts of inventory. An ideal type of system is something like the pneumatic tubes used outside of teller windows at banks. Assembly material handlers could therefore send for material in small quantities as they needed it. What is not needed is large fork trucks to move small parts in small quantities. They require aisle space when the objective is to close the spaces between machines, and there is always the urge to fill them up.

Making every work center operate as if it were part of final assembly means close attention to its location and the methods of moving material to and from it. Not only must the layout be tight, it must be set up so that material does not have to be double-handled. All this points back to the need for housekeeping. Well-organized stockpoints contribute to the quick identification and movement of the correct material. Compacting the plant layout has objectives other than just freeing space for other uses; it reduces transport distance and improves visibility for coordination so as to attain balance.

Balancing the operations in a plant is a dynamic thing. Unless effort is made to keep production running at the rate at which operations are balanced, after a time production needs to be rebalanced at a different rate. The fear of losing line balance is the major underlying reason why, traditionally, there is reluctance to stop or slow a balanced assembly line when it is running. This is also true of an entire plant. There are two countermoves for this:

Visibility and signaling systems to permit everything to respond to temporary variations in overall parts usage rate, and to changes in planned mix.

Keeping a little slack at each work center to allow flexibility to respond to deviations from the planned schedule. The example in Exhibit 6–4 has very little slack to respond to deviations from the routine. It needs a little more, perhaps 20–30 minutes per shift.

These measures allow flexibility to get back on schedule if a plant drifts off, but flexibility is a tool for overcoming problems of a temporary nature. The objective is to hit the schedule rate precisely if possible. Most assembly lines are monitored to see if they are holding a steady pace. Each hour, a posting shows whether the number of units which should have been completed up to that time have actually been completed. The objective here is linear production, hitting the production target hour after hour.

This seems to contradict the practice of stopping the line to make corrections if something goes wrong. If good process control has been developed, most of them last less than a minute. They are usually the result of someone fumbling a tool or a part, getting a screw cross-threaded, for example. They need a few extra seconds to avoid having a missed operation. The line resumes speed in less than a minute. Such line stops are easily absorbed in the allowance for flexibility. Major problems are a different matter. Those require overtime to correct, and they usually demand follow-up action to make permanent changes which will prevent some recurrence. The major problems usually reflect bad quality parts, machine breakdown, or some problem which the system should not permit to happen.

Since the schedule changes periodically, the operations of a plant may need to be rebalanced periodically. By experience, operations usually can continue without rebalance if the cycle times required at various work centers do not vary by more than ±10 percent; but beyond that, work center organizations may need change. The detailed planning for production must take this into account, the subject of Chapter 10.

Balance is also important in bringing in material from suppliers, but that is the subject of Chapter 9. It is very important inside the plant, and can be summarized in several characteristics of plant management using stockless production:

Operations are set to be more people-dependent than machine-dependent. If equipment is kept simple, this is not an expensive policy, and it is a necessary one for flexibility.

Workers need to be multifunctional in the sense that within departments they can run different positions as the need for rebalancing occurs. There is great emphasis on building up the total span of expertise of each worker. A house full of specialists leads to great difficulty adjusting balance or schedules so as to keep all these specialized positions working.

Operations need to be rebalanced as they are improved, and also from time to time as schedules call for major changes in cycle times. This means a large volume of rebalancing work. Most ordinary schedule rebalancing should be within the capability of supervisors. (Industrial training programs for stockless production companies spend much time on line balancing and operations balancing.)

Operations that are fully automated still need to be rebalanced from time to time. That puts the major emphasis in equipment selection on flexibility—fast setup and a range of effective running speeds.

Material handling patterns and timing need great attention to also make the movement of material quick and flexible. The best way is by simplifying so that people move material in small quantities over short distances, followed by the ultimate of one-piece-at-a-time conveyance rapidly by simple mechanical transfer devices. The plant must avoid choking on its own material-handling methods.

The layouts are compact, for short transport distances and easy visibility.

Capacity planning in the short run is a matter of adjusting work hours and of rebalancing operations, usually more the adjustment of work hours because that is easier. (Japanese plants on stockless production typically work a nine-hour day. If the schedule drops, they shorten the work day, and if it increases, they lengthen it or work Saturdays—a practice with its counterpart in every industrial nation.)

Capacity planning in the long run means having the equipment to be as people-dependent as possible in the short run, and also to having the capacity for both proper maintenance and for flexibility in responding to problems. Flexibility means the ability to revise and rebalance operations.

PREVENTIVE MAINTENANCE

Preventive maintenance is built into the stockless production system. The key to preventive maintenance is repair and adjustment *before* problems occur. The key objectives:

1. Reduce downtime from all sources—make processes operable any time they are needed.

2. Reduce variation in performance:
   a. Eliminate special adjustments—"chewing gum and bailing wire"—when setting up or to keep it running.
   b. Eliminate equipment, as much as possible, as a cause of defects by keeping the operating tolerances of equipment within a narrow range.
3. Extend the life of equipment.
4. Prevent major equipment repairs.

Reason number 4 is the one usually given the most attention in support of preventive maintenance, and that leads to controversy over its value. Managers must then check whether preventive maintenance is worth it in comparison with keeping the equipment running until it needs major repair. With stockless production, the first two reasons are vital. The major benefits of the system cannot be realized without them, which makes preventive maintenance get the respect it deserves.

Preventive maintenance is everyone's business. It begins by integrating preventive maintenance time into the production schedules. One of the best ways to do this is to allow time between shifts for preventive maintenance (PM), and the usual pattern is for plants to work two shifts with a two- or three-hour gap between them. That gives time for both preventive maintenance and for overtime work if they are needed for the schedule.

Responsibility for PM becomes part of every operator's job, but this cannot be done by urging them to watch the equipment more closely. It begins by making the correct operation of equipment at all times a matter of urgency so that attention is given to any change in noise or performance pattern, just as would be true if production equipment were commercial aircraft. Everyone understands that aircraft should have regular tire changes, replacement of hydraulic cylinders, and replacement of parts before wear causes malfunction. The key is for production people to feel that their equipment should likewise be reliable, and that no equipment is too inconsequential for maintenance.

Operator responsibility begins with housekeeping. Wiping oil, cleaning rust, and painting are psychological because they promote the attitude that only good-looking equipment should be used, but it also requires the operator to pay some degree of attention to equipment while cleaning. It also prevents poor performance or malfunction for unexpected reasons—low oil levels, warning lights not working, jams from chips, and the like. It is also important for the operator to set up the machine if at all possible. It creates more pride in the machine, and it requires closer observation than if the operator merely watches it run. This also helps dispel the propensity for equipment abuse—the feeling that "I can operate it as I please, and someone else will figure out if something is wrong. The company has more than enough money for repairs anyway." (The author has watched operators deliberately jam machines just to take a break while someone else repaired them. That might not have happened had the operators felt that a machine breakdown was *their* responsibility and breakdowns caused *them* more work.)

The level schedule that comes with stockless production is very conducive to a regular PM schedule. One of the reasons for abandonment of PM programs is that equipment is not made available for PM at regular times, so the maintenance people assigned to PM find themselves skipping through a planned schedule trying to find an open machine and rescheduling others for later. If the time for it is not made available and PM people begin to have idle time, the management may conclude that it will not work, and the PM people themselves soon understand that they are assigned a low-priority job. With a level schedule with slack time built in, it should not be difficult to incorporate PM on a daily, weekly, and monthly basis. If operators are responsible for seeing that equipment operates properly, they will assist in seeing that maintenance specialists perform the PM they cannot do themselves. The best situation is for operators to discuss the equipment with maintenance personnel at the time of the PM work—perhaps even assist them where possible. Here is another area where multifunctional worker training pays off, though there may be some obvious work rule customs to resolve.

Stockless production practices provide a number of ideas to improve PM practice:

A modification of the card system can be used for tools (attachments if they are used on setups with machines) as shown in Exhibit 6–5. The card system keeps control of tools, especially a record of breakage and damage. Each card authorizes the attachment to move between just one work center and the tool and die shop. It provides traceability as to what is happening to them. Operators may even sign the cards denoting tool use. That keeps a running record of how much the tool was used and who used it. It is simple, quick, and does much to not only place responsibility for tool damage but also promote process control through consistency of tool use and refurbishment.

Keep a record with major attachments such as dies. A simple card in a jacket will do. Recording the number of hits on a die can eliminate guessing about die wear. Special problems can be noted. It can also hold any special instructions about adjustment of a particular die for setup. The most obvious problem is for operators to remember to keep the record. It helps if they feel some responsibility for the length of die life.

EXHIBIT 6–5 Move card system for tools (attachments)

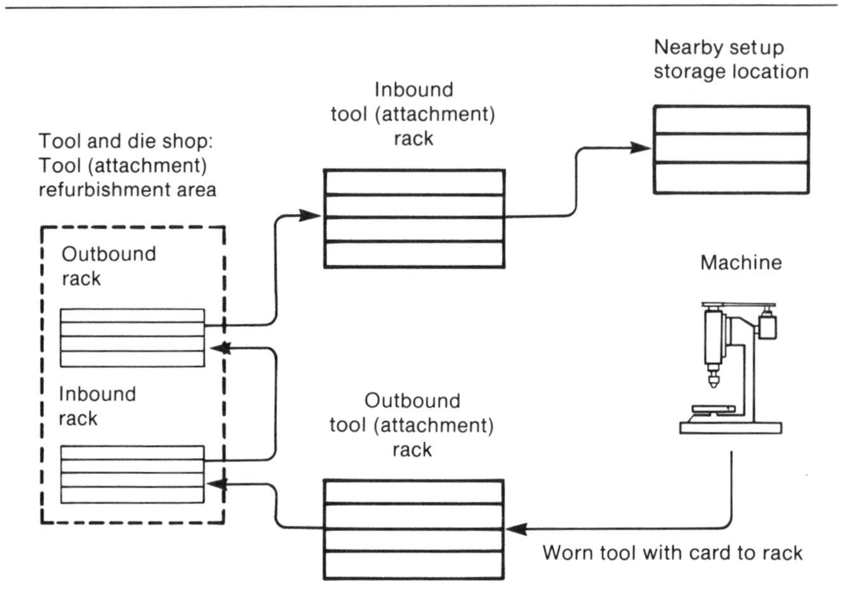

Case 1: Used tool is to be thrown away. Take tool and card to tool and die shop. Leave tool (attachment). Remove card. Use it as authorization to get a new tool from the outbound rack.

Case 2: Used tool is to be refurbished. Take used tool and card to tool and die shop. Card may stay with used tool as traveling record of performance. Select another tool with card from outbound rack. Total number of cards for each type of tool keeps upper limit on the inventory of that type of tool.

Incorporate automatic warnings to check tool wear. For example, when is it time to sharpen a drill bit? Set a counter on the drill when it is set up. Then the counter can trigger a warning light when the number of hits indicate that a drill bit is probably worn. For this to work, one has to set up the machine so that it will start up only after the counter has been set. Another variation of the same idea is to trip an indicator that keeps a record of how many times a machine has been set up with a particular drill bit, if it is expected to last through several setups. In order to work, this type of system must be very quick and self-policing. It saves extra time in checking workpieces or drilling times to see if a bit is worn. The idea can be used for a number of applications—recalibrations, lubrications, and so forth.

Keep a simple record of machine breakdowns and their causes. One of the simplest is to place a sticker on the machine each time it breaks down. Then anyone can see which are the problem machines. An extension of this is to keep a card in a jacket attached to the machine itself. On the card, record the times and causes of breakdowns. The system has to be very simple because it depends on operator diligence at a time of stress.

The usual problem with preventive maintenance is that it depends on people knowing that it is important. Like many other things, this depends on management's attitude toward it. One of the major steps in many plants is to stop making the emergency repairs a heroic act while everything else is dull routine. The way to do this is by placing great emphasis on equipment modification. A breakdown should be followed by modification to prevent recurrence, for which maintenance personnel and operators are responsible. This means assigning some of the most skilled maintenance personnel to do modification, not assigning them to the next breakdown. It only happens if management recognizes those who work to prevent the recurrence of breakdowns and does not give all the accolades to those who make a habit of doing only the minimum job necessary to make equipment run again. (Exhibit 6–6 shows an overview of a PM organization for stockless production.)

This is one of the most difficult steps in maturing to stockless production. Without a major effort to upgrade the PM of equipment in order to prevent breakdown, stockless production sputters along because major sources of problems do not receive the attention required in order to stop them. Eventually, the problems will be covered with inventory to make life comfortable again. The review of the details of equipment performance is vital, not only to improve equipment reliability, but to gain process control. Without dedication to this, management wants to have the rewards without paying the price, and the price is not in money but in management resolve.

VISIBILITY SYSTEMS

There is an old story, which is told in several forms, about a restaurant owner having trouble between the waitresses and the short-order cooks. The waitresses bawled the orders verbally to the cooks, who did not always get them correctly and did not always prepare them in the order in which received. The problem intensified at the busiest period, the lunch-hour rush. Waitresses blamed cooks, cooks blamed waitresses, and the customers were understandably unhappy. As the story goes, the restaurateur sought advice from several management consultants. One suggested an order code system, but that did not help much. Another suggested hiring female cooks since males did not like to take orders from females. That was impractical for lack of

EXHIBIT 6–6 Departmental responsibility for preventive maintenance

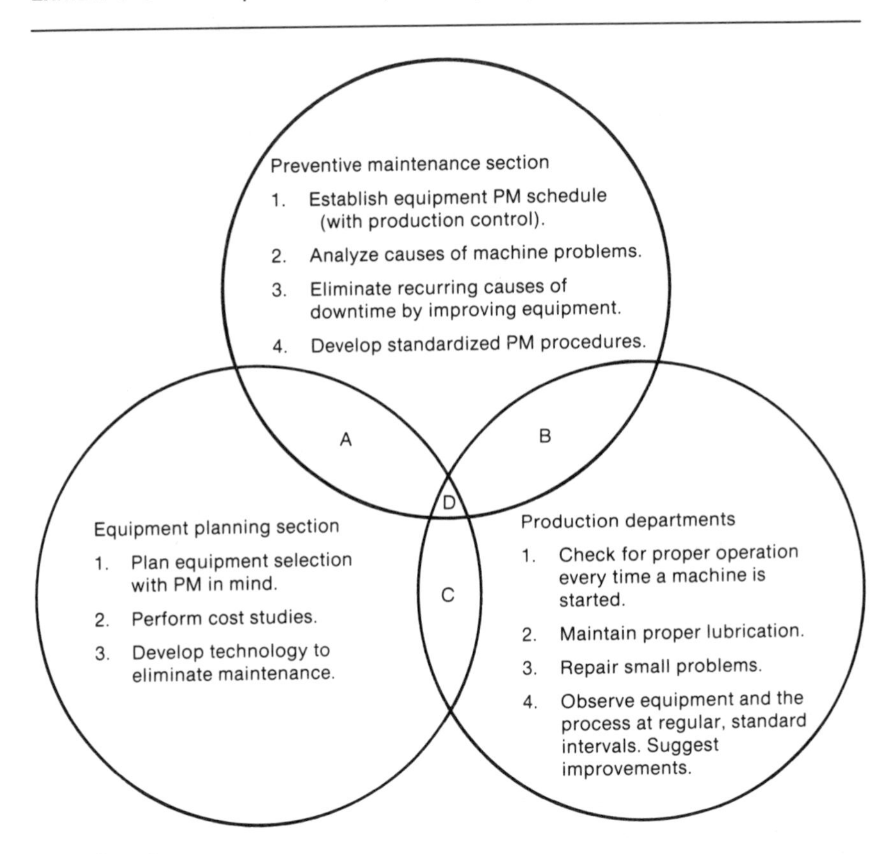

Interface A:
1. Develop maintenance procedures for new equipment and new operations.
2. Supply PM data and machine histories to equipment planning section.

Interface B:
1. Supply PM data and machine histories to production departments.
2. Instruct operators in proper methods to operate equipment to avoid breakdown and damage.
3. Production workers record minor problems and observations for the preventive maintenance section

Interface C:
1. Equipment planning provides operating instructions to production.
2. Production department provides records of equipment and process productivity to the equipment planning section.

Interface D:
3. Evaluate the total PM program.

This exhibit is based on a modified description by Isuzu Motors of their PM organization.

female cooks, or any cook who would stay more than two weeks. One suggested that the owner pass the orders from the waitresses to the cooks and take the responsibility for it himself.

The solution turned out to be the ordinary turnstile method for hanging written tickets so they could be rotated from waitresses to cooks. With only a small amount of extra walking, the waitresses gave written orders to the cooks in the proper sequence. This principle of rapid communication through a simple but formal system is very important in any situation where many people must communicate accurately in a hurry. It reappears over and over again in production. As the pace of material movement quickens between work centers buffered by little inventory, it is very important to stockless production. High visibility systems for group communication are embedded in most of the methods, and it applies to many things besides material flow.

The card system itself is a method of highly visible, rapid communication. The cards themselves may become cumbersome to handle in some situations, and some substitute for this method of signaling used. Electronic signals are a possibility. However, where these are used, the most important principle is visibility. The objective is to communicate, not just to make the circuits in a computer work. The signal should arrest the attention of humans, and with stockless production it is better to overdo the visibility than to have confusion. There is a tendency to feel silly when introducing things like brightly colored washers as tokens for the card system, but they work. Rapid, accurate communication is essential for maintaining a balanced flow of parts and tooling.

Most stockless production plants have a large board prominently located. Japanese call it an *andon.* This board gives the status of every work center in the plant according to a lighting system. In the simplest case, a red light comes on whenever a work center is stopped. This allows everyone to adjust to trouble, both at the work centers they feed and at the ones feeding them. The andon signals may be simple or elaborate.

Most andons also show how many final assembly units have been completed so far during a day versus the number that were planned. That lets everyone know where they stand. They may display the cycle rates of various models in final assembly. That presents the "drum beat" of production.

As equipment becomes more and more automated, a system of lights on the machines is important to allow a few workers to monitor which machines need attention. The lights need to be big enough to be seen from a distance, not just little lights on a control panel which can only be seen from one direction. One of the common systems for these lights is the following:

Green: Running normally—go condition.
Yellow: Needs a tool change or adjustment.
Blue: A setup change is necessary or imminent.
White: Time to take a quality control sample or recalibrate a gauge.
Red: A malfunction stop. (Comes on if a quality measurement stops the machine.)

A part of housekeeping is to keep all locations marked: part locations, tool locations, instruction locations—mark where everything is if it is frequently needed in a hurry. Test gauge and test jig locations are prominently marked.

Safety warnings are common in all industry. To stop presses or other equipment, light beams which can be broken by a hand are preferred. Using the two-button method (both hands must be on the buttons for a machine to be activated) is not considered a good use of the operator who has to stand and watch the machine work. In other instances, a decibel meter will activate a flashing light if noise becomes loud enough that ear protection should be worn. Lights, bells, whistles, and mirrors are all part of this. There is considerable study of what is useful for gaining human attention. It only needs to be used, but used with judgment so as to avoid giving people information overload.

This extends to high visibility on the status of people and projects. With multifunctional workers, a qualifications board in each department is very common. This shows which workers are qualified on which machines or for other tasks.

Suggestions get high visibility. They are commonly posted on departmental boards. Supervisory responses are also posted so that everyone can track the progress made in implementing suggestions.

The most important point about visibility systems is that they are essential for teamwork in fast-moving production. They allow everyone to participate as a team by reacting to signals without delay. One of the benefits of stockless production is the plant control that people have without excessive communication problems, which cause mistakes and delay. Inventory covers the communication problems, which are among the "rocks" and are broken up by visibility systems.

WORKING INTO FULL AUTOMATION

Stockless production has the goal of eventually reaching full automation, but doing so only as technical progress permits it to economically become the way to attain the most compact, straight flow of material. Automation just to avoid human processing of workpieces is not a goal. A new stage in automation is developed as soon as it becomes clear that further progress in efficiency by manual means is not possible. Automated equipment should not subject material to more waste and delay than people do.

Automation is an imprecise term. Different people mean different things by it. Some mean that workpieces are processed by a machine without operator positioning or assistance. Others mean that workpieces move through an entire plant untouched by human hands. However, everyone means that physical actions formerly done by a person can be done by a machine, so one way to make the definition of various stages of automation more precise is by detailing the different degrees of interaction of man with machine, as shown in Exhibit 6–6.

The definitions in Exhibit 6–7 are adapted from a similar table prepared by Shigeo Shingo, but they follow no generally accepted American definition, except to reflect how terms appear to be commonly used. However, it is not possible to assuredly define the roles of operator and machine in a general case; the definitions are only in reference to a specific task performed with specific equipment.

NC and CNC equipment is generally programmable to move a tool or workpiece through a set of paths and speeds to perform a task. Much of it is used for machining operations. It generally cannot pick up a workpiece and replace it, but must be, at best, bulk-load so that it can feed parts along a limited path for processing. Such a machine can be programmed by an operator on the spot or, if linked to a computer, can call a program from memory for a new setup. An operator must generally make the attachment substitutions necessary for a setup.

A robot is more flexible. The Robotics Institute of America defines it as a "reprogrammable, multifunctional manipulator designed to move material, parts, tools or specialized devices through variable programmed motions for a variety of tasks." It too may be reprogrammed on the spot, or it can generally recall a new program from a computer memory. Humans usually need to make some physical changes for a setup, but because of its flexibility, the robot has more capability to set up itself. There are several classifications of robot:

1. *Pick-and-place robots.* These may be defined only by their task of moving a part or tool from one location to another. To do this they may need mobility in no more than two planes and only simple rotational capability, though the job sometimes calls for more complex motion and positioning capability. Many pieces of equipment performing pick-and-place operations would not be called robots.
2. *Servorobots.* These generally employ hydraulics and accurate positioning sensors so that they have three-dimensional mobility

EXHIBIT 6–7 Operator-machine interaction at different levels of automation

Action	*Manual operation*	*Semiautomatic machine*	*Automatic machine*	*Programmed* automatic machine*	*Robot*	*Linked automatic machines*
Setup	Operator	Operator	Operator	Operator or machine	Operator and machine	Operator and machine
Load	Operator	Operator	Operator-bulkload; machine–piece-load	Operator-bulkload; machine–piece-load	Machine	Machine
Run	Operator	Machine	Machine	Machine	Machine	Machine
Unload	Operator	Operator	Machine	Machine	Machine	Machine
Transport away	Operator	Operator	Operator	Operator	Operator	Machine
Other	Operator	Operator	Operator	Operator	Operator**	Operator**
Program	N/A	N/A	N/A	Operator	Operator or machine**	Operator or machine**

* Numerically controlled or computer controlled (NC or CNC).
** Adaptive robots may detect deviations or sense new tasks and reprogram themselves.

and are capable of arcs and full rotation. There are some subcategories which fall within this:

a. Programmable robot. A designation which usually means that the robot is programmed by microcomputer and a memory contained within itself, or by a dedicated controller. It can often be reprogrammed by leading the arm and gripper through a new sequence.

b. Computerized robot. A designation generally meaning that the robot is linked to a larger computer which allows a variety of programs to be stored in memory and recalled on command.

c. Sensory robot. This type has a programmed learning capability which allows it to reprogram or reposition itself in response to sensed stimuli. Example: A robot which can follow a freehand chalk line for cutting steel. This type is still in early development for practical application.

Other types of robots are designated by the tasks for which they were programmed—for example, assembly robots or welding robots. However, robots have captured our imaginations because of their potential flexibility and their association with tasks considered exotic. The cost of the computer equipment associated with robots has been decreasing, making them practical candidates for many tasks formerly done by people.

Proceeding from simple automation to more complex, programmed automation is a journey which demands that managers study exactly what automation is supposed to accomplish. A factory filled with linked automation demands exacting guidance. Small matters that never before needed conscious attention become important. For example, robots need to know exact locations from which to select parts. They must discharge parts to exact locations for automatic pickup. They require that material be formed in the exact way called for by their program. For example, a welding robot just follows a programmed pattern. It does not compensate for imperfections in the edges of the pieces to be joined. Adaptive robots which can deviate in a simple way from a programmed path are expensive, but even if they were not, do we really want them to deviate? Too much deviation will cause scrap which jams the system. Robots and other programmed equipment may be slaves in one sense, but in another, they are taskmasters which require people to think through what needs to be done and why.

People working on the shop floor make mistakes, but many of them never become serious because humans adapt—sometimes by making offsetting errors. This brain work under unanticipated conditions makes man difficult to copy as an instrument of work. Exhibit 6–7 was prepared to detail the point that the functions of work most easily done by machine are the repetitive ones, and those least easily done by machine are the nonrepetitive ones that require adaptation. Non-adaptive automation requires a precision in plant layout, workplace organization, dimensions of material, and uniformity of work pattern that remains unseen with human workers. Developing these conditions so that better tools can be used is another way to state the objective of stockless production.

In specific cases, we must study the interaction required between man and machine. The proper role of people is most important. It even has to do with esthetics. Works of craftsmanship are prized because the craftsman leaves a record of his struggle with imperfection in each piece, and many of these records show the individual craftsman's own style of prevailing in these struggles.

People are adapters, consistently seeking new experience, continuously reacting to changed conditions, some much more so than others. That is stimulating. Human development comes from creating and remembering ways to overcome a variety of problems. On the other hand, repetitive work can almost be done while asleep. People doing repetitive tasks find other things to occupy their minds as a primary stimulus while going through the motions.

This leads to two seemingly contradictory rules of process design in order to bring out the best in both people and equipment:

1. Never continue having a person perform a job, or a part of a job, which a machine can do better, if it is at all economical to do so.
2. Never advance the degree of automation of a job without first exhausting the potential for improvement by human means.

A shorthand way of saying this is, "Routine for the machine and development for the human:" People are the developers and controllers of the machines. The machines should be used in such a way that people remain alert. For example, an operator should always have something to do other than watch the machine. It is how we use what we have that is important in the production process, and if both people and equipment are developed so as to employ the best attributes of each, an excellent system will follow.

Developing automation through stockless production means three things:

1. Development of the system to move material as quickly and accurately as possible from raw material to finished product. If that aim is accomplished, whether tasks are performed manually or automatically is secondary.
2. Proper preparation for the advancement to the next higher stages of machine automation by previewing what the problems will be in practice. Or, thought of somewhat differently, work through

incremental improvements rather than make radical engineering changes and spend years in debugging them.
3. Maintenance of the proper roles for people and machines so that automation is not degrading to people.

The argument can be made that if an automated industry is the goal of stockless production, then why not plunge ahead with professional engineering and develop it. If engineers know enough about the production of the product to do that, that is a correct decision, but much of the know-how in many production processes is in people's heads or their "little black books." The process must be understood in detail and be in statistical control.

Builders of automated factories can anticipate some of these problems by sheer engineering analysis, but creating totally automated factories requires the linking of a huge amount of hardware and software, a project that is never totally debugged and that always requires substantial human intervention for maintenance. It is not a simple matter of mechanizing and programming the obvious. We must constantly study the nature of the production process itself.

This begins in the design of the product. The nature of the tooling, equipment, and work force skill are a factor in design because nothing can be built better than the process is capable of building it. Robots and other automated equipment are expensive and complex, but they are still tools, and their characteristics must be learned by humans so that we know how to apply them skillfully. A man may unconsciouslessly compensate for flaws in material which an automated machine is oblivious to. Tolerances are only what the equipment design and programming can deliver. Evolving into the use of automation requires revising the interaction of workers, machines, and material which make up the production process, and that requires learning.

Once it begins to mature, stockless production becomes a philosophy of debugging the way to automation. Start by improving the reliability and flexibility of current equipment through modification and preventive maintenance. Then balance the system so that the equipment can be further modified to operate in synchronization. Introduce a different design of equipment into the process only when progress has reached a point where the modification of old equipment can no longer be accomplished. At that point, we should know much more precisely what kind of equipment is needed, so we do not need to introduce—and pay for—equipment with much more capability than is really needed.

How big the increments toward automation should be is a matter of judgment. The experience of those in stockless production is that increments can be very small. In the beginning, no one knows exactly what the problems are, so the production process must be studied and problems incrementally found. Keep finding and solving small problems, and, after a while, many large problems will be reduced to where they can be solved as small increments also.

Chapter 1 mentioned one of the most famous plants developed to full automation by this philosophy, the Kamigo plant of Toyota Motor Company, really a complex of engine plants, which, as a whole, produces several sizes of engine. One of those plants produces six-cylinder engines, 1,500 engines per day with two shifts and 161 total employees. One afternoon the author observed 48 people working—28 in final assembly, 9 in test, 3 at material handling, and the rest roaming through several acres of equipment, keeping it tended.

The Kamigo plant is highly automated, but there are no robots, and none of the equipment is programmed by advanced methods. The equipment appears to be 1950s-vintage equipment, all modified and connected by transfer lines and devices developed in-house. The sensors and signaling system employ very little that cannot be purchased in an ordinary hardware store. This plant was developed by stockless production, and it reached a state of full automation about 1966.

Block castings enter the plant fresh from the foundry and emerge as finished engines 6½ hours later. From pouring the block until it is mounted in a vehicle averages 36 hours. Physically crude it may be, but no offloading and reloading interrupt the progress of material through the Kamigo plant. The defects in the system are minor because it was developed by very carefully applying what was available to the needs of the production process—no technological overkill.

Linking all this equipment together required modification to balance the cycle rates on all machines. Since the cycle rates may vary from one schedule period to another, the cycle rates on each machine should be easily changed from one rate to another, elementary on some equipment, but not all. For example, impact speeds and pressures are important variables in a forging operation. Forging may only be possible within a limited range of hit speeds and pressures if the workpiece is to be free of distortions, so the actual work rate of the impact may be determined by the nature of the process. Other types of equipment may likewise allow no variation in the work stroke itself.

Given this condition, the most effective way to balance the overall cycle rates of many different machines is to modify them so that they completely stop for different periods of time between cycles. Using cams, microswitches, and similar devices, many types of equipment can be synchronized to the same cycle rate by the following four-point method:

1. At the completion of each working cycle, stop automatically at the point where all components and attachments of the machine have returned to the start position.

2. Sense whether there is space to hold another finished workpiece on the outbound transfer line or device. If so, automatically eject the workpiece to be transported to the next machine. If not, wait until space is available and eject.
3. Sense if an unprocessed workpiece is available from inbound transfer. If so, feed this workpiece automatically, using guides to position it without human assistance.
4. Sense when the fresh workpiece is in position and automatically start the next working cycle.

If the work cycle on every machine is faster than the cycle rate demanded by the schedule, this method will connect a great deal of garden-variety production equipment together to feed an assembly line. The system is a mechanical version of a pull system. When all the accumulator space in transfer lines or devices are full, each machine makes another piece only when it senses that there is a space to put it. If the assembly line stops, the accumulators rapidly choke with material, and there is no space to load more material into the gateway machines.

Another important principle is to keep the transfer lines short and the accumulators small. The reason for large accumulators is to buffer delays between machines in case of breakdowns or defective production. Large accumulators are a sign that the root causes of problems have not been overcome. Long transfer lines, which serve as large accumulators of inventory, mask problems in the same way.

Keeping it simple is important. With the four-point plan, most equipment can be modified and maintained by the skilled trades people who work in most plants. It does not require a tangle of wire or a labyrinth of microprocessors. It is also unnecessary to hire a lot of specially trained technicians. If the modification is done by the people already employed, they develop themselves as they develop the equipment. As they develop, more programmable complexity can be added. The level of complexity at which a plant starts depends upon the skills of the people now there to master and modify it. (We cannot get into the computer age any faster than the ability of people to master the skills and use them knowledgeably to do those things necessary to move material from the raw state to finished product without waste.)

EQUIPMENT SELECTION

The selection of equipment needs to fit into a plan for evolving into full automation without waste:

1. Buy equipment which is flexible or which can be made that way, and which can be modified and maintained in-house.

2. Do not box in a department or operation from development by buying equipment which cannot be moved or linked, unless there is no other reasonable choice.
3. The quality and reliability of the output are important.

An obsession with payoff of a piece of equipment can defeat this. In the worst case, this can lead to getting a piece of equipment with a high potential output rate while giving too little emphasis to anything else. This equipment is not going to run any faster than the cycle rate demanded of it, so high speed will be a detriment if it cannot be fitted into a balanced pattern with other equipment. It is the overall effect of all the equipment together that counts.

When selecting or modifying equipment, setting up U-lines, or just organizing equipment into balanced work centers, it is important not to be deceived by inappropriate measures of equipment utilization or by wrongful measures of efficiency. If a piece of equipment operates on a 30-second work cycle for three weeks with a high efficiency rating, but has to be shut down for a week to work off the inventory, it might as well be given a 40-second production cycle during which it can perform a 30-second work cycle. The overall output is the same, but now it is balanced with the rate at which output is required. Confine the idle time to the smallest increments possible.

Likewise, buying "expensive toys" is not necessarily progress. Equipment which has many features and capabilities not really needed may or may not be easy to modify for the present purpose. Example: using a robot with full multidimensional capability to do a pick-and-place operation which requires only two-dimensional capability. The job could be done by a simple swing arm, and the money would be better spent developing other parts of the production process.

In fact, from a financial point of view the ideal automated factory or automated industry has a very low break-even point. In the case of equipment in the U-line, one can see that the controlling factor in equipment utilization is the manning of the line. The equipment is idle much of the time. If it were expensive, that approach could not be afforded. The fully automated plant is analogous. If it can be avoided, we do not wish to create a monster demanding to be fed with a high level of production in order to pay for itself. In that sense, we wish to operate equipment above capacity, but that means operating above the capacity at which it will financially support itself. We do not wish to operate it above its intended speeds prescribed by kinematics or the control system.

There are many exceptions. For example, the initial use of robots almost everywhere has been to replace humans in jobs that are hot, dirty, noisy, or dangerous. That also allows people to become acquainted with robots and their use before trying to skillfully fit them into a pattern of automation that includes many other types of equipment.

Much of the disappointment that companies experienced early in the history of NC and CNC equipment came from buying machines which people did not know how to use skillfully, which had a long setup time, or which were difficult to program. Therefore they discovered little work of the volume or profitability necessary to justify putting on the machine, so many pieces of NC equipment gathered dust. They became useful as people learned enough about them to regard them as tools. Then they knew what they really needed, and when and how to use them.

These kinds of issues become buried if managers try to boil down equipment decisions to purely financial cases to compare against hurdle payoff rates in capital budgeting. To make progress, equipment decisions must be made according to a stockless production strategy, not according to criteria which are blind to it.

7

Quality systems in stockless production

Stockless production is dedicated to improving quality, but improving all the other aspects of production is not at the expense of quality. The system will not work without the elimination of scrap and rejects (or, for some processes where such elimination is not attainable, it will not work without achieving high and stable yields so that pull signals can be served with dependable production). Many points have been made in the preceding chapters about quality problems as a major source of waste and delay, and the purpose of this chapter is to concentrate on stockless production as a system for quality improvement. Most of the ideas are useful even if a company is not able to develop a pull system.

Some quality defects, such as a missing part, are easily defined, but that is not true of all defects. Quality is often a hard thing to define. Just because a part fits and works is not a sufficient indicator of quality. It can fit and work and still fail to function very long, or it can be unsightly. Quality implies functionality, durability, reliability, and pleasing appearance. Some of that comes from the design of the product, but what the customer uses and sees comes from the production process.

Conformance to design specifications is often stated as the criterion of acceptable quality from a production process. That assumes both that the specifications have been determined to assure quality performance and that no problems of measurement obscure whether the parts really do conform to specification. Many arguments in inspection attest that these are not assumptions to take for granted. If designers are unsure what tolerances will be held by production, they may set them a little tight, meaning a closer tolerance range than is really necessary for the part to perform its function over a lifetime. They may require the part or assembly to pass a test that is arguably irrelevant to the excellence of product operation in any conceivable environment. If the actual level of process control in production makes it difficult to pass such tests, inspection may become "loose" in order to get out production. If this happens, the criteria of conformance are lost in the fog over what is really necessary to make the product work satisfactorily, and if the situation degenerates into irresolution, the actual output quality degenerates into something unsatisfactory.

Specifications and tests of conformance are always important, but a quality system built entirely on them tends to become one that relies entirely on inspection—sorting out the defects, not stopping them until the scrap rate becomes unbearable. Even then the tendency is only to take the minimum corrective action to bring output back into conformance with the specifications, if they are known to be operationally important. Through this approach, many managers are tricked into thinking that there must be a cost trade-off to improve quality. Stopping production temporarily to correct defects decreases apparent production efficiency as measured in the short run. It seldom costs to improve process control over the long run because that eliminates the need for rework or unplanned downtime to correct deficiencies.

Every part and every product is a result of the quality of the manufacturing process which made them, so a program to permanently improve quality through improved process control must consider quality defined in two ways:

1. The quality of the part or product itself.
2. The quality of the process which made the part or product.

If the production process cannot be controlled well enough to make a product that meets or exceeds what is expected of it in use, it makes little difference what the design specifications were. If the production process is well within the control range needed to make the product, there is no reason to have anything defective leave the plant. Quality is built in, and a quality production process must be fail-safe so that people with learned skill but ordinary talent can create excellent products without exception. If the production process itself is always on the verge of going out of control, they do not have the means to do this. Just tightening inspection and telling people to be more careful is of minimal and temporary help.

Some of the defects do come from worker errors. Dr. Edward Deming is fond of quoting from his experience that only 15 percent of the specific causes of variation come from the workers, however. The other 85 percent come from the system because it is set up to allow defects. The objective is to change the system so that the equipment should always produce parts within specification and the workers will catch their own mistakes, or, better yet, be prompted so as not to make them. This is a very different type of activity from doing only the minimum necessary to avoid what is currently thought to be an unacceptable level of defects. There is no such thing as an acceptable level of defects, and the program should strive to constantly improve the statistical control and fail-safe methods of production to prevent defects from being made. TDK Electronics Company refers to this as "beyond zero defects."

QUALITY ACTION: PERMANENT IMPROVEMENT OF PROCESS CONTROL

The first version of the heading above read, "Corrective Action: Tightening the Process Control." It had a nice ring to it. It might have attracted a second glance from readers saturated with hyperbole, but it would have been misleading. *Corrective action* implies an adjustment to bring into control something that was once under control, but even if parts are being made and assembled within specification, there is no way, without knowing why we are doing something right, to be sure that this condition will continue. *Tightening the process control* implies a temporary action, decreasing the allowed ranges of adjustment for equipment perhaps, but without necessarily continuing the program.

It is very hard to keep in mind that quality comes from an unending search for improved process control, every day doing something to make it a little better. Only a firm management persists in this. Many myths grow up to justify relaxing, all of them with some plausibility, and all with the undesirable potential of distracting people from systematic improvement. A few of the myths:

1. *People do not want to do good work.* There are many variations of this, and it is true that some workers are careless, but they will never improve as a group without a program which concentrates on the means of improvement.
2. *There is a cost trade-off between quality and productivity, and we cannot afford to reach a point of perfect quality.* Poor quality also costs, but the cost may not be immediate or obvious. It

may not show up in the costs captured by the formal cost system. Any defective is costly because time and material have been expended for something which is of no value. This really means that we do not see how to improve our process control without vast expenditures. It does take development of skill. Almost anyone can improve quality of the output by increasing cost, but it takes skill to improve quality while decreasing cost.

3. *We have an excellent quality control department.* A few knowledgeable people do not make a quality product. They cannot convert a production process from a condition of noncontrol to one which is in control if others are consistently neglecting to take action. In that case they can only produce parts fit for use.
4. *We have quality control circles.* This is a more recent variation, but the quality circles can only raise the level of consciousness about quality unless those in them know what process control means and go about making improvements to attain it.
5. *Our competitors have the same problems.* This is a variation on the cost trade-off excuse. It works until a competitor no longer has the same problems.
6. *Our problems are different.* Everybody's problems are somewhat different; the *pattern* of problems is remarkably similar from one manufacturer to another. We have to work through our own problems. No one will present specific solutions on a platter.

There is seemingly no end to the excuses that can be made, nor to the "preaching" which can be addressed to the excuses, but all of this only defers becoming serious about the fundamentals of process control. Difficult it may be, but there is no substitute for management demonstrating the importance of process control by building consideration for it into every production decision.

IMPROVING PROCESS CONTROL BY IMPROVING PROCESS CAPABILITY

Variance is a measure of the spread or dispersion in the distribution of results. It can be defined statistically as the average of the squared deviations from the mean of a set of measurements. Zero variance indicates that every measurement is exactly the same. The idea can be extended to measure deviation from what was expected as a standard.

Variation in a final product can come from thousands of sources if the product is complex, and there can be multiple sources of variations in even simple parts. Tooling, material homogeneity, equipment variation, operator variation—all influence it. By concentrating on the complexity of the statistical variance which could come from many sources, we can persuade ourselves that process control is too complex a matter to really achieve in mortal life.

The principle of attack is simple. Reduce variance in any aspect of the operations where it can be reduced. *Do* something wherever it can be done with diligent attention and affordable cost. By reducing the variance from small causes, ways to economically reduce variance from more difficult sources may become more clear. The process will finally come under control when the critical variables of a finished part *never* should fall outside an acceptable range for a random reason. Then whenever a part is checked and any property is out of tolerance, there must be a reason, an assignable cause which can be found rather quickly.

The principle of this approach is very simple but very important in practice. The statistical and technical complexity of a specific problem in process control can be very formidable if there are many sources of variance, some of them unknown. Attacking it as a complex problem can be very time-consuming and costly. Correcting what is simple to correct means reducing variance wherever it is found until the amount of variance left is reduced and the sources of it become clear.

Much of the basic concept is explained by the problem of stacking tolerances, usually a part of the apprentice training of machinists. The idea is depicted in Exhibit 7–1, the point of which is that design tolerances desired and the variances obtained from production processes are two separate things. The variance in Exhibit 7–1 is shown as a statistical normal distribution. If the variance in production is too great, design tolerances cannot be obtained and production variances combine statistically.

With a little statistical knowledge in mind, many people can become conscious of the ways their every action can introduce unnecessary variance into the production process. The objective in production is to reduce as much overall variance as possible from the process until we get the result shown in Exhibit 7–2.

The example of process control in Exhibit 7–2 assumes one variable on only one part. The interaction between variables is a consideration, and the number of sources of variance can be very large, but the variance in the result is always the sum of the variances of the process which creates the part.

Variance in the part = Variance in the machine operation itself,
+ Variance in tool dimensions,
+ Variance in setup adjustment,
+ Variance in material homogeneity,
+ Variance from operator control,
+ Variance from contaminants present,
+ Variance from all the other possible sources,
some unknown.

EXHIBIT 7–1 The problem of stacked tolerances

Part design tolerances expressed like this. . . .

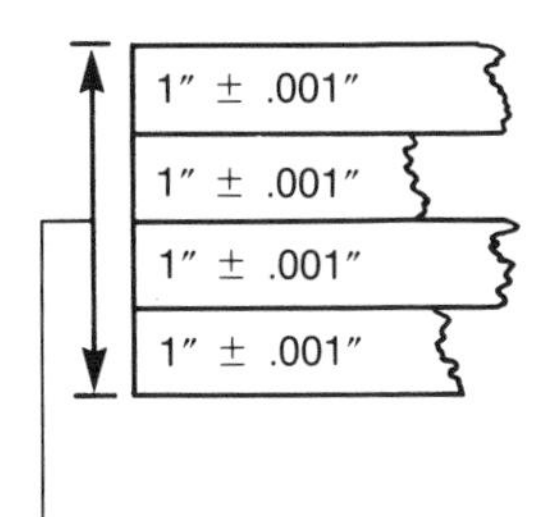

The sum of this stack will *not* be 4″ ± .001.″ Oversizes and undersizes of the separate parts will sometimes combine into an extreme total variation.

May be attempted from a production process with a variance like this. . . .

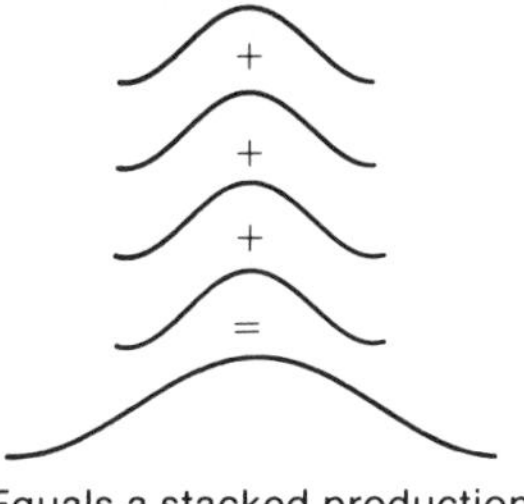

Equals a stacked production result like this. . . .

Or better, from a production process with a variance like this. . . .

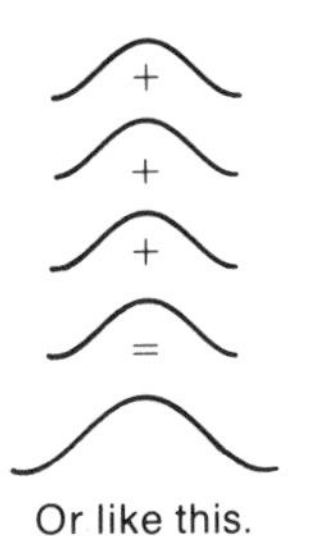

Or like this.

It is very helpful to describe the variance statistically, but that is not the objective although it may help our understanding of the process and its problems. The objective is to reduce the variance (however described) as much as is feasible to do so.

This explanation of process control is basic. It is certainly elementary to quality control professionals. However, it is often neglected in manufacturing practice because the attention is given to only doing the minimum to pass inspection rather than improving the process so that defects are eliminated.

The purpose of bringing each process under statistical control is to make 100 percent automatic inspection useful for making corrections and improvements. If the range of controllability of the process is narrowed enough, then, for practical purposes, the probability of even one part being made outside the accepted range for "normal" reasons is zero. Then, if even one part outside this range is found, it becomes feasible to shut down the process, find the cause, and correct it.

This approach applies to all production which lies well within the state of the art. If the production task is to machine a three-foot-diameter titanium ring to within 10-thousandths of an inch, it taxes the limits of technology. Process control in the sense described is a rational goal when the technology is not pushing the limits of the possible, so enhancement of standard production equipment and procedures will

EXHIBIT 7–2 Reducing variance to obtain control of a process

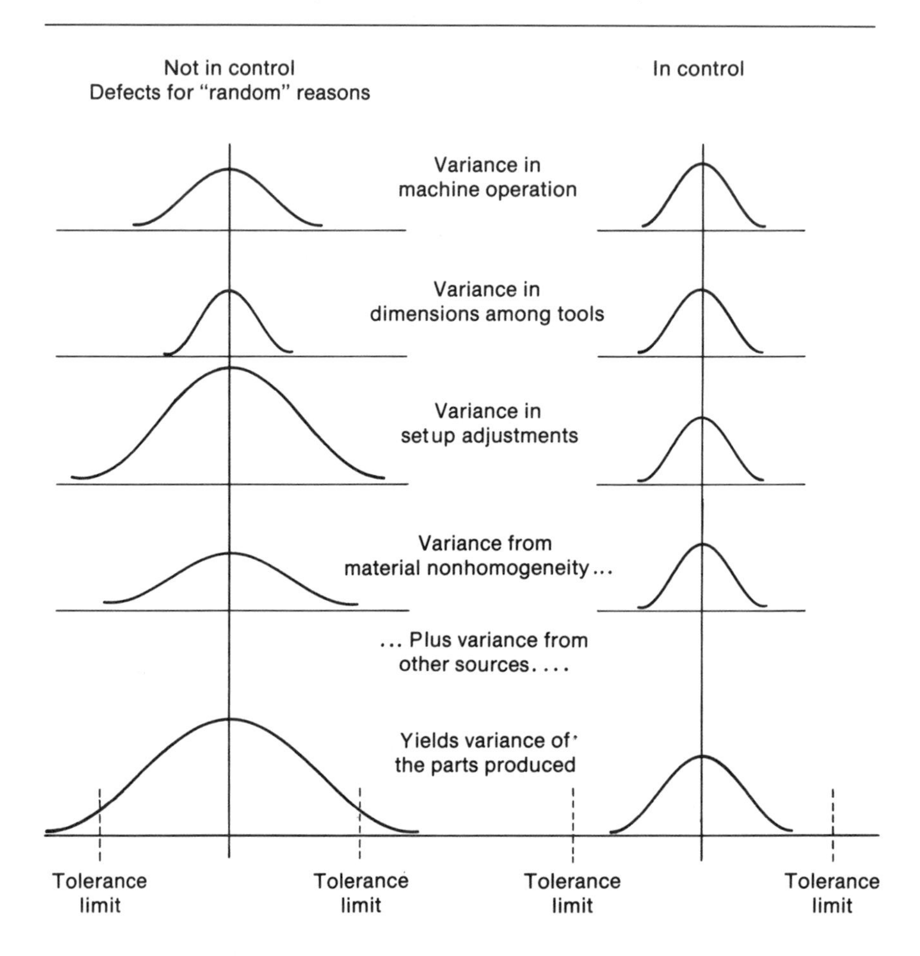

accomplish it. Several types of activity in production stem from this idea of narrowing the range of variance by eliminating much of it.

Inspecting or checking the variance of the process, not the tolerance of the part. For this to work, the variance capability of the process must be well within the tolerance limits required of the parts. If the basic process is capable of this, then inspection consists of making sure that it stays that way, so the purpose of inspection is to be sure that the process has stayed in control, not to see if we have produced an unusable defect, although that is obviously important if it occurs. Then the design of the inspection process depends on the nature of the process, not just the requirements of the part.

Once most production processes start to drift out of the basic variance range of which they are capable, they keep on doing it, usually for a reason which can be determined and corrected. If production runs are short, wear and various factors which normally cause a process to drift out of control are of limited duration. Therefore if the first production piece indicates the process is in control and the last piece indicates it is still in control, the changes are about 100 percent that every piece in between was in control. The setup was the same, the source of the material was the same, and overall conditions were the same. This is true for many common processes—stamping, molding, forging, die-casting, and so on.

That makes the inspection process for most stockless production runs simple. Check the piece to see if it is within the dimensional limits (and other physical standards) which should be held if the process is in control. The sample size is two, first piece and last piece. Perform the inspection at the machine immediately after the part is produced. If the process is out of control, then corrective action is needed. If the last part is the one out of limits, but just barely out, the parts may still be usable, but we have to pay attention to the process before the next setup. However, if the process is carefully tended to, the chances are that whatever happened is more catastrophic than that, so corrective action is taken and replacement parts produced.

In any case, unusable parts are not passed on to the next process in production unless there is a breakdown in the diligence applied to the system. Once the capability of the process is established for it, the procedure is so simple that there is little excuse for that.

The true use of a control chart is to establish the basic limits of the capability of a process and check whether output is staying within those limits. It is used as needed as a matter of daily routine to check whether wear or some other condition is causing the process to drift outside the limits of which it is capable. In a less routine way, it is used to suggest improvements whereby the limits of process capability can be further narrowed.

If a control chart is in use when one is trying to find ways to reduce the total variance of a process, the control limits may be set at ± 1 standard deviation, not ± 2 or ± 3, as is commonly the case. The idea is to examine the process at those times to seek ways to reduce the basic variance in it.

Of course, if attempts at refining the process fail, inspection comes down to sorting out the unusable parts. If the percentage of them is high, output may come to a complete stop if the sorting is done at the machine itself, but doing exactly that as soon as possible is what is desired. It puts the emphasis on examining the process, not inspecting the parts for sorting; and examining and controlling the process is what leads to permanently abandoning delayed inspection. Very little progress can be made with stockless production as long as that is the prevailing practice.

Fail-safe production and inspection. Other kinds of quality problems exist besides those of assuring that every part is within an accepted tolerance on dimensions or other properties. Some are easy to see, but seem to be chronic. Examples: missing or cross-threaded screws, loose trim, gaps in seals, tension on electrical wire (so it pulls loose from connections), loose hose clamps. There is an obligation to see that operations are correct and complete. Whenever feasible, this should be done by 100 percent automatic inspection because the diligence of neither human operators nor human inspectors is likely to assure this otherwise.

If automatic inspection cannot take place immediately after an operation is complete, it should take place automatically when the workpiece is positioned for the next operation. For example, if a screw is missing, a limit switch on the next setup jig will fail to depress. Unless the use of jigs, limit switches, or wiring complicates the equipment so much that it slows the quick setup procedures, this is preferred because it prevents double-handling or delays just for inspection. Some of the sensors can be built so that they are no hindrance to normal setup or operation at all; for example, setting a welding robot so it shuts down if there is a blip in the current, indicating a gap or a glob in the weld produced. A classification of methods is in Exhibit 7–3.

If near-perfect production of nonexotic products is expected, mere mortals cannot attain zero defects unassisted. (The author cannot even find all the typos in the manuscript.) The production process has to be designed to assure that all oversights of consequence are detected and corrected as soon as possible.

Every little bit of assistance with even the trivial operations helps. For example, if an operator is to put eight screws in a panel without fail, it helps to do it against a lighted background if possible, so that there should be no pinholes showing at the completion of the task. If in addition, eight screws are doled out at a time in groups, that is a numerical check, and the practice, trivial as it is, is consistent with the material-handling ideas expressed in Chapter 6. Then if the subsequent operator or a check jig with switches looks for the presence of all the screws, one is down to a very low probability of oversight. Overkill? Not if it is important to appearance and function that this be done without fail. By such oversights is the rework bay at the end of assembly lines filled. A rhythm for checking is vital for assuring that processes are constantly monitored. (See Exhibit 7–4).

The effort should go into stopping the causes of a problem, not into correcting the results of a problem. The causes of the problem should be found and corrected anyway. The rest is unnecessary work, and it

EXHIBIT 7–3 Classification of fail-safe techniques

Class	*Method*	*Description*
3	Signs or signals to warn or remind operator and thus prevent mistakes	Visual or audio effects prevent or detect mistakes. Examples: Differentiate similar parts by color code. Operation must cover a background color, stop pinholes of light, or change in sound, and so on when complete or correct.
2	Jig method	Jigs used for production operations are modified, or inspection jigs are created to prevent or detect mistakes by one-touch inspection: Detects installation of wrong component or wrong orientation of component, Detects incorrect dimensions of part. Does not operate when part is incorrect or an assembly error is present.
	Machine method	Replace human operations by mechanical ones to eliminate human inattention or error.
1	Automation	Limit switches stop the machine or the line if there is an irregular occurrence.
	Automatic inspection	Inspection instruments are modified to warn of defects or irregularities by *two* or more signals (Light *plus* buzzer, or light *plus* air jet). Part or product locks into jig when defective.
	Positive orientation or "cassette" design	The design of a part itself prevents assembly error because it only fits one way.

Each class of technique has its advantages and some expense of installation. By ingenuity, we may be able to use as many of these techniques as possible without great expense or time lost. The class 1 methods should be used for all critical parts and assemblies.

This was adapted from a classification by Isuzu Motors. Other companies have similar classifications.

EXHIBIT 7–4 Rhythm method for quality control checking

If operations can be performed to a drumbeat of production, then a pattern of quality control actions can also be done by a rhythm, thereby minimizing the chances of oversight.

Objectives

1. Define areas of quality control, checking responsibility clearly.
2. By rhythmic checks, prevent defects from leaving each area (often defined as narrowly as one operator's scope of control.)
3. By rhythmic checks, prevent accepting defects into each area.
4. Propagate the method throughout all production operations.

Method

A pattern of quality control checks and inspections suitable for each department and area must be determined by each supervisor, sometimes with advice from the quality control professionals. Establish different control actions to take place at specific times or frequencies, such as:

1. Start or end of each day.
2. Start or end of each shift.
3. Every setup, or every change of material used.
4. Every *n*th piece.
5. At beginning or ending of break periods.
6. By regular signals such as a bell or chime.

Examples

1. Check first and last piece when changing setup.
2. Check first piece when changing roll stock feeding a press.
3. Plot a sample on a control chart for a machining operation every time a signal is heard.
4. Check a part and extruder and die conditions every time a bell rings.

Measurements that are recorded at specific times and intervals can usually be combined for statistical analysis, and trends or drifts can be easily detected. The system builds quality checking into the operating routine of all parts of production. It is the kind of quality inspection pattern usually desired by quality professionals and it is extended into the fabric of production activity itself.

plugs the flow of material even longer. Rooting out these kinds of things is a major part of attaining linear production, and, incidentally, the count of units completed should be based not on how many have finished the nominal process on the line, but on how many have cleared the area with no remaining touch-up to attend to.

Most of the effort is not in understanding how this is supposed to work, but in overcoming the habit of covering for anticipated irregularities in the production process. The objective of stockless production is to put the maximum concentration on making the primary process itself fail-safe, not on complicating the total process by covering its flaws with backup procedures. People have to work their way into this state of mind by gaining confidence in their equipment and in themselves.

As confidence grows, process control becomes more and more the *anticipation* of potential causes of defects and of making improvements before they occur. If a defect occurs, stop the process and correct it, but the objective is to anticipate problems, so as to avoid stopping the process, while still assuring that every piece is a good one. Even if everything is going well and defects are rare, we should still strive to reduce the number of ways a defect could potentially occur, consistently reducing the randomness in the process—that is, progress from control by trial and error in the "black art stage" to control by a sure knowledge of cause and effect.[1]

Inspection, checking, and immediate feedback. If we inspect and recheck every part as it is produced, the flow of production will slow and the cost of the inspecting will soon defeat the objective of better quality at lower cost. With common sense we can usually determine which operations really need special attention to bring them into control and keep them that way. Of the ways to simplify all this, the practice of immediate feedback is the one which contributes the most.

Many kinds of defects are easy to see or simple to correct. If something goes wrong, the part will not work in a subsequent process. If the time between when a part is made and when it is used is short, then straight feedback from the using operation may be all that is necessary, especially if the production of the part can be quickly and easily done. In the U-line for example, if the dimensions of a part are a little off from one operation, it will not fit properly in the jig for the next operation. No special checking is needed. If production is fast moving, the same kind of principle applies without U-lines.

It is a very obvious thing to do on an assembly line. Operators can usually tell if they are being fed incomplete or incorrect work from another station. Obvious as it is, the practice is not always to go straight to the source and inform them of the problem if that is not the management style on that line.

Immediate feedback is an integral part of the quality systems of stockless production. It should become the management style throughout the production network because it allows quality work to flow smoothly by eliminating the unnecessary complications.

How expensive and difficult is it? A normal finding is that preventive maintenance and attention to equipment and its attachments will do a great deal to improve process control, and sometimes it improves it to a point where all defects are nonrandom. This is not expensive. What becomes expensive is if the accuracy of all the dies and tooling must be refined. That can cost a great deal and for many processes it is

---

[1] In Japan, the practice of stopping the line whenever a defect occurs or is about to occur is called *jidoka*. According to Fujio Cho, production control manager of Toyota Motor Company, "Make problems visible to everyone, and stop the line if a problem occurs. All thoughts, methods, and tools to avoid line stops are Jidoka." It is the anticipation of problems that is really the end objective.

prohibitive. In that case, we *may* attain the effect of process control by regular sorting of the output. The most common applications are those in which matched sets of components can be sorted for assembly:

Rollers and races for roller bearings.

Piston rings and cylinder bores.

Sorted resistors for balancing circuits.

The typical finding is that there are not many situations that demand sorting and that a program like this begins to pay for itself very quickly because it greatly reduces defect levels and eliminates the *hidden plant,* a term sometimes used to designate all the rework areas and nonstandard operations that have been set up to work out the problems resulting from lack of process control in the originating operations. If that is not expected to be the case, the company should review whether its products tax standard production methods to their limits, or whether it needs to review design criteria in view of what is really possible if production were enhanced to its full potential.

Summary of developing process control

These methods are basic. This discussion has omitted measurement accuracy, statistical design of experiments, different designs of control charts, runs tests, control theory, and other analytical tools within the kit of the professional quality control specialist. Getting control of the more complex production processes will likely require application of some of these. However, the majority of process control problems stem from not applying the basics. Devising ways for a total check of missing operations seldom requires advanced analysis.

The importance of process control provides a strong argument for keeping production processes as simple as possible. It is easier to establish process control for simple technology, but simple technology is not necessarily old technology. It is easily controlled technology—the avoidance of making every step in production so loaded with potential sources of variance that it is difficult to sort them out.

If process capability is attained, a stoppage of the process on detection of a single defect is adequate control, but better yet is an automatic warning that a process is about to drift out of control. Knowing what to do to overcome or prevent problems is what is really desired—back to the issue of using the quality control process to trigger action on the basis of our detailed knowledge of cause and effect. Knowledge of the most basic statistics of variance sensitizes operators to why problems may occur and how serious they are. Knowing what specific corrective and preventive *actions* to take converts this sensitivity to practical ends.

Process control is a part of the fabric of stockless production. Inventory and lead times are not reduced without process control; in turn, reducing inventory and lead times stimulates the detailed actions needed to obtain process control, operation by operation. Like the other aspects of stockless production, it is really learned by doing. A brief summary of other actions for obtaining process control and process capability follows.

1. Reduce the randomness of conversion processes:
   a. Refurbish and maintain equipment so that it performs within a known range of variance.
   b. Develop a regular pattern for reworking attachments to avoid defects because of wear or drift from adjustment. (Note that this implies a reduction in the variance of the work schedule, another way to think of reducing variance in the production process.)
   c. Develop standardized ways to refurbish tools and attachments so these have consistent characteristics. (Two different tool-and-die workers do not always rework the same tool the same way, for example.)
   d. Set up the equipment regularly in order to develop skill in knowing correct adjustments and in avoiding the onset of non-controlled conditions. (This is the opposite of letting equipment run until something *has* to be done.)
   e. Use material which has consistent properties; do not mix grades, types, and sources for the same part. (This is one of the strongest arguments for buying material from a single source—the same physical source, not just the same company. For example, try to always run steel from the same rolling mill through the same press if material consistency is necessary for control.)
2. Always take actions that tend to narrow the variance of output, not the reverse.
3. Set up 100 percent testing with feedback as soon as possible after each part is made. After a process becomes free of random defects, the source of every defect can be tracked. Shut the process down immediately to look for it.
   a. If possible, keep testing and feedback automatic. The part should be measured the instant after the making, and the process automatically stopped if there is a defect.
   b. If this is not possible, try to automatically inspect as part of the following operation, or perhaps workpieces can accumulate through two or three operations and be autoinspected at a fourth. Trap defects as close to their source as possible.
   c. It is important to stop for correction of the cause of a defect should one occur, provided the cause is readily determinable,

not random. This prevents the process from degenerating by causing discomfort to the operators, so that they must correct the source of the problem or the machine will continue to stop. It puts a premium on quickly diagnosing and correcting problems.

d. Set up 100 percent automatic inspection wherever possible to check for completeness—missing or misaligned parts. All humans lapse on occasion, so make the process fail-safe.

e. If inspection cannot be built into a production operation setup, design it to be done with as little added labor and time as possible. Example: Place the subassembly on a check jig as part of conveyance to the next station.

f. Avoid having people do nothing but inspection. To the maximum extent possible have those who do the work responsible for inspecting the work with the inspection built into the production operation as much as possible.

g. Reduce inventory between operations as much as possible. This reduces delay time as much as possible. If a part does not work properly in a subsequent operation, immediately inform the source operation and make correction.

h. Do not rely only on urging of the operators to remain alert. That works for short periods, but not year in and year out. The process itself must be designed to prompt operators to make correction when needed.

The last point deserves elaboration. If quality is to be built in, the management must constantly stress what the *work force has accomplished,* not what the management has accomplished. In a highly automated facility, correction of problems requires a high level of teamwork so that every worker supports every other worker without hesitation whenever required. For example, if a worker pushes the red button to stop an assembly line, within seconds co-workers should converge on the spot. They give assistance without discussion if it is apparent what is needed. Most such line stops occur in Japan because an operator is in danger of allowing an operation to go by incomplete. A screw has cross-threaded or a piece of wire broke—small but important details in the correct assembly of the product. Correction of such problems occurs quickly, and many line stops last less than 10 seconds because of instant double coverage of the problem. The idea is exactly the same as baseball players backstopping each other on fielding plays. A process stop must be a call to action, not an excuse for a break.

More than a matter of culture is involved. As operations become more automated and they become more closely tied together, more teamwork is required to maintain process control. If operators are in charge of a large amount of automated equipment, they must cover each other as problems occur. The signaling systems and training are designed to bring that about.

Recognition of responsibility for quality is a very important part of this. Accepting responsibility for quality comes when management demonstrates that there is recognition of people for doing so. For example, at Kawasaki USA, the defect rate on one assembly line dropped about fivefold within weeks as soon as the foreman and his workers understood that spotting and correcting problems in assembly was their responsibility: Every unit should finish assembly in such a condition that workers are "willing to put their name on it and hand it to the customer at the end of the line." That began to build the teamwork. Since they should not send the problem to assembly repair at the end of the line, they began to look for the causes of problems themselves.

CAUSE AND EFFECT: MAKING THE PLANT INTO A PRODUCTION LABORATORY

Most manufacturers actively seek the causes of production problems only when starting up production or when the occurrence of defects threatens the continuance of production. They have not established systematic ways to ferret out and eliminate causes of defects and other production problems. Many engineers get their plant experience as an occasional troubleshooter, and sometimes that is an unwelcome diversion from other activity. Few people are charged with daily responsibility to consistently establish the specific causes of existing and potential problems and work on them, so most production people are self-taught in the science of cause and effect in production.

Scientific or development laboratory experience is helpful, but factories are a different environment. There are usually more people involved, and most of them consider their job to be "getting it out," not experimenting. Experimental design in laboratories is more prestigious than "shaking down" a plant, so there are few articles and books on the latter activity while there are many on scientific experimental design, but the logic for identifying the causes of problems is very similar. The problem is controlling the environment. The objective of an experiment in a laboratory is to control the environment and observe results. Turning a plant into an environment more like a laboratory is a process of learning to control conditions, an expansion of process control to repeal Murphy's Laws wherever found.

Converting a plant to a production laboratory is a management program to consistently seek and refine knowledge of the production process, even when there are no serious problems at the moment. It is very important to remember that the program covers not just quality in the *product,* but quality in the *process.* If management is only satisfied with good enough, they become accustomed to problem-solving by whatever level of daily crisis is necessary to do this. The need is for rigorous use of simple methods by which production people can observe and correct problems. There are three principles for a program:

1. Keep problems visible. Do not permit the kind of practices to continue whereby people hide their problems.
2. Develop methods of organizing information so that people can draw conclusions from it.
3. Control the environment so that sources of variance are not a permanent jungle obscuring what is really happening.

Keep everyone aware of the state of product quality—and process quality

Everyone, both staff and operators, needs to be aware of the specifics of the state of product and process quality. Signs and exhortations that "quality is our most important product" are not sufficient. Stockless production is intrinsically filled with the kind of immediate feedback practices which help a great deal, but there are more ideas:

1. Keep the fabrication work centers in as close contact with the final product as possible. If everyone can see how the result of their work fits into the overall product, they will have not only psychological incentive but also daily observation of what is really needed to make the process work smoothly. In effect, every worker will know that what goes to the customer has his or her stamp on it somewhere.
2. Provide specific information on quality feedback from the field to everyone. The layout of the results of a teardown analysis from a warranty claim is excellent feedback. The argument against this is that such open displays will spread the word to the marketplace that the company's products are defective and adversely affect sales. However, that is hiding the problems when what is wanted is to eliminate them. If feedback is positive rather than negative, display that also, but tell the truth about the specifics of quality. Excessively protecting people from embarrassment is the enemy of a serious quality program.
3. Keep running records of defect rates, and always peg a better performance as a goal. Satisfaction is another enemy of quality improvement. When the defect rate has reached less than one in a million, switch emphasis to making the functionality or appearance of the product better—a higher standard of improvement.
4. Keep a running record of significant line stops or operation stops and their cause, even if no defects were actually produced.

Organize records so as to classify problems and point to causes

This is a basic, everyday thing to do. One can swamp production people with requests to keep information, so the supervisor needs to keep a watch on it. Better to keep a few items of information religiously when these point to problems than to keep much data that points to nothing. People respond to the problems on which they must report while others may remain unseen. Basic as it is, this area of management deserves regular review.

Classify defects and other occurrences:
- —Pareto charts classify defects by type or by source, with the most frequent category first.
- —Histograms and runs charts classify by time, or they may be used to classify in other ways. Both Pareto charts and histograms can dramatize raw numbers.
- —Production logs are also important, but they are often kept in a haphazard way that does not convey potential sources of problems at all. It is helpful to provide the logs, if kept, with headings that promote regular rational observation and action. For example:

Observed condition	*Assumed cause*	*Corrective action*	*Observed result*

- —The objective is to turn the production log from something that merely shows raw data, from which efficiencies are calculated, into something which stimulates the urge to make improvements—something like a laboratory notebook.

Change what is being recorded as process control improves. There is no stimulus if new problem areas are not attacked. There is negative stimulus from filling out forms if the data do not lead to something. (There are obvious exceptions for those items which must be kept for permanent record.)

Use methods for tracing defects or other problems to their sources

Many tracing programs are a temporary response to problems, but a few may be permanent. Stockless production enhances the ability to trace by emphasizing fixed routings and streamlined flows. Tracing is used to identify the source of a part by machine, by operator, by type of material used, by supplier, or by time of day made. Tracing helps establish how material is moving through a flow as well as identify the source of defects. There are many ideas for tracing:

Specially marked containers may suffice to trace material back to a specific source—a particular machine or operator.

Use colored stickers or paint spots.

Apply stickers to parts that indicate date, time, location, or other information pertinent to the part history. (Tachikawa Spring Company applies such labels with the same kind of labeling gun used for price stickers in groceries.)

Bar-coded labels can be scanned to reference computer records. The potential of these is enormous for identifying material once it has escaped the confines of the containers used for production. It takes a capability of generating and printing labels to make up bar codes for unique sources, but a company that has gone to the general use of them should have this capability.

Various kinds of tracer markings have application for tracing the routes of tools for denoting machine attachments suspected of malfunction. This leaves a highly visible record of the frequency of malfunction of various pieces of equipment and other situations, useful when establishing sources of specific variance for process control.

Make detailed comparisons of operations

Most people who have done troubleshooting of production problems have noted that what they thought was happening in production turned out later not to really be occurring. Looking in a procedures manual to see how operators or maintenance personnel are performing specific tasks may be very misleading. Differences in results come from different operators, different machines, different setups, and different maintenance procedures, to name only a few. A direct way to check for uniformity of production is to compare different operations. Industrial engineers frequently find a variety of methods in use for the same task. These differences may or may not be of consequence in process control. A methods study of different operators or machines is one way to compare, but a very good approach is to have meetings between different operators. Meetings not only permit the staff to become aware of differences affecting process control, but they also make the operators themselves aware of the differences.

A common approach for comparing operations is to have meetings of different work crews:

Cross-operation meetings. These meetings are between two different work groups doing the same thing at two different work centers. For example, Hewlett-Packard Company holds cross-line meetings between assemblers building the same product on two different assembly lines.

Cross-shift meetings. Compare organized notes on how different operators use the same equipment and handle the same parts. These also help clear the air over issues that normally create acrimony between people who work the same area on different shifts.

Cross-functional meetings. Compare notes on how work in remote locations fits together. For example, hold meetings between operators and those in the toolroom who repair the tools. That helps explain a great deal about breakage and nonstandardization.

Such meetings accomplish little if they ramble, so operators or others who compare their work must come prepared. They need to describe in detail how they perform various tasks. That means that they need a little basic instruction in industrial engineering methods analysis, and that does them no harm. The meetings are very useful if their purpose is to fill in the gaps in knowledge while trying to further process control.

The leaders of such meetings must also be imbued with a sense of the purpose of each meeting, and see to it that participants are prepared. If participants are not prepared to discuss and compare operation details, such meetings can degenerate to blame-casting. It also takes patience, because the first few meetings may not reflect much skill on the part of the participants, and that must be developed.

Organize a cause-and-effect analysis

In analyzing problems in many factories, the investigator is never correctly aware of the conditions prevailing at all times. What an engineer or supervisor thinks operators or machines are doing, based upon limited observation, is not taking place consistently. Most of us have experienced this from time to time with automotive problems. A fault observed in driving refuses to replicate itself when the vehicle is hooked up to test monitors. In a factory, it may be a mystery what happens at 10:30 P.M. with the second shift operator.

Likewise, some machines, or attachments for them, seem to have characteristics all their own. Material may seem to be homogeneous, but an occasional discontinuity is difficult to repeat for careful observation. Many production problems seem to have this quality of now-you-see-it and now-you-do-not. Diagnosing the problem takes repeated investigation.

What is needed are ways to gather and organize the observations of many people, and also to communicate among them what has been tried with or without success in search of a solution.

If a problem really gets confusing, several people may be making changes in operating conditions at once without properly communicating. Sometimes the problem goes away without anyone understanding the key change that caused it to disappear. That leaves uneasiness because no one knows what may cause it to reappear without warning.

Without getting into a statistical design of an experiment, the garden variety method for almost all production problem solving is to hold everything constant except one variable and observe the result. For example, if parts stamped on a press are dimensionally out of tolerance, one might keep the press setup untouched but change to another roll of strip stock which has been found to have less variation in gauge and observe the result. If that happens to work, the resulting action is to specify strip stock with less gauge variation, but unless this becomes part of the permanent history of the process, someone else will introduce "bad" steel into the press six months later after all has been forgotten. Too often, such matters are carried only in people's heads, and heads full of experience may not always be available. Even when committed to memos, the knowledge may be lost over time with the annual cleaning of files.

From time to time, most plants attempt to document what is required for proper operation of processes by updating procedures manuals. Regular attention to this is a big help, but procedures manuals do not always explain why given procedures were originally recommended, and conditions change over time.

This approach usually bogs down because a stimulus for ongoing study of each operation is not maintained. The current crisis gets attention. Other areas receive neglect. Procedures manuals, or their equivalent, should perform two functions:

1. Maintain a status of the operation so that people understand why all procedures are used.
2. Provide a basis for quickly entering into cause-and-effect studies, either in response to new problems of parts being unacceptable or for further refinement of process control.

It is this nitty-gritty level of organizational memory that provides the underpinning for processes being always in control, so they deserve to be treated with respect.

The basic way to organize a cause-and-effect program of improvement is illustrated by the cause-and-effect diagram. The basic form of the diagram shows the relationship of various potential activities and conditions on the property to be controlled. In using this logic, one should be careful to precisely describe what it is that is to be controlled or improved. Then many people may be able to contribute observations which might have some effect, much like taking testimony from witnesses to an accident.

However, the basic cause-and-effect diagram, or fishbone diagram, as it is sometimes called, is a personal tool to organize the logic of attacking a problem. To make it a coordinative tool, it needs to be enhanced. One way is by setting it up as a communications medium with the addition of cards, as shown in Exhibit 7–5. Setting up such a diagram is a matter of formalizing the logic to be used in approaching a problem, and that in itself adds clarity to thinking.

One person should maintain the diagram with cards. The supervisor or the engineer who is principle investigator is the logical choice. The key is to keep the cards near the site of the problem so that all operators and others can write observations on a card and add it to the collection at any time. Arranging the cards on a board in a place of easy access promotes this.

Many variations of the cause-and-effect diagram are possible. The important point is communication. A notebook kept by the machine may have almost the same effect, but only if the notebook is efficiently filled out by all those involved with the problems. The same may be true of the diagram. It is necessary to find a medium of communication that expresses, in a way visible to all, the current status of progress in reducing or in keeping to a minimum a particular variance in the production process.

Note that the cause-and-effect diagram is not just a method for collecting thoughts and suggestions for problem-solving. It can be used for problem-anticipation and -prevention; and as process control advances, that is the superior use. It illustrates the kind of logic needed for vigorous and persistent pursuit of process control in production.

There are many ways to approach this. The cause-and-effect diagram with cards has been very effective at Sumitomo Electric Industries. It was a central portion of the methods of improvement which they used to greatly improve process control. Dr. Ryuji Fukuda wrote about it in a paper which won the Nikkai Prize from the Deming Award Committee in 1978.

Most important is to constantly have a method of presenting and communicating the state of production technology so that everyone daily can become aware of the current state of process control, and can proceed from there to make further development. There is no dishonor in attempting to make an improvement that does not succeed, but the company can only slip if there is neglect to constantly improve anything which can be improved.

Nothing has here been presented beyond a combination of basic methods of scientific investigation with basic methods of good communication. These can be woven together in an infinite variety of patterns which fit the situation. There is very little to it which we do not know. However, we are strongly subject to knowing much but doing little.

EXHIBIT 7–5 Organizing a cause-and-effect analysis

Basic cause and effect diagram

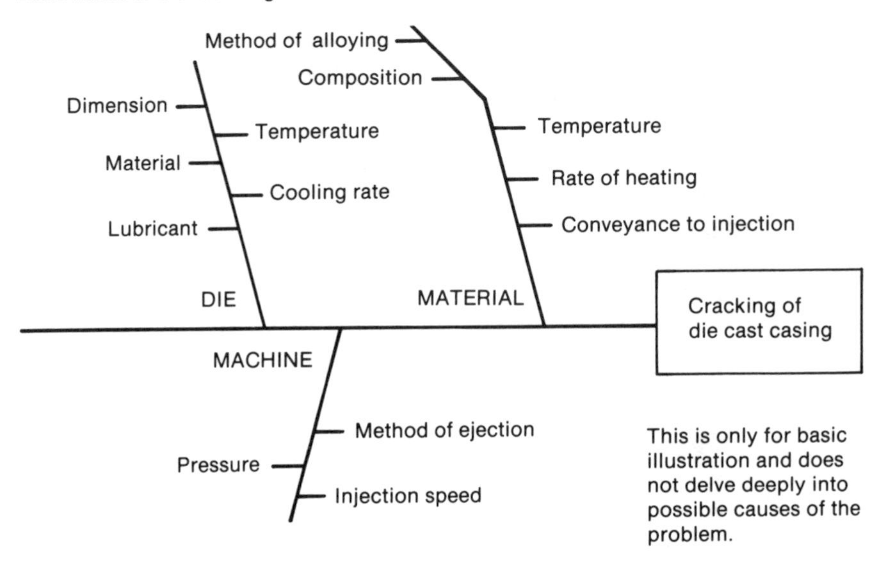

Cause and effect diagram with the addition of cards

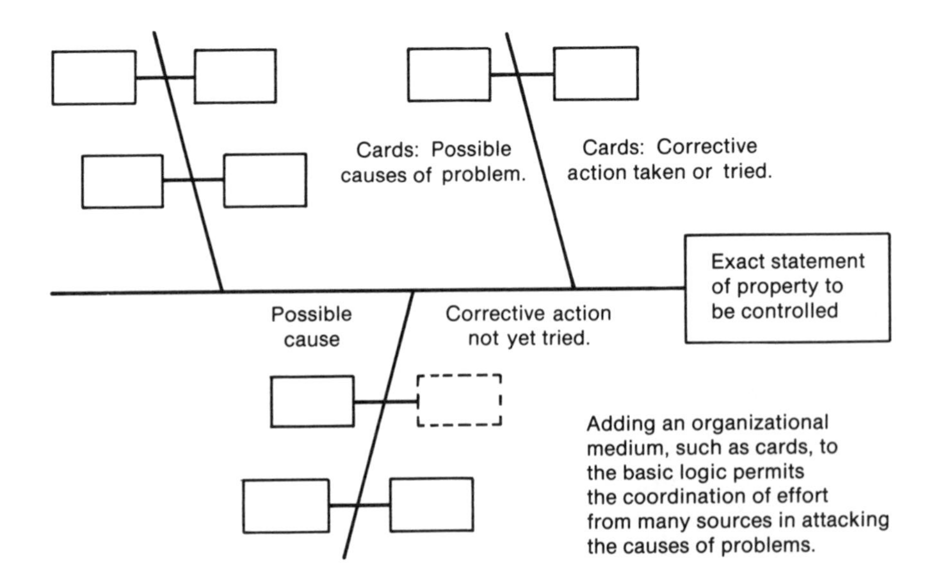

Experimenting: The surest way to build knowledge of cause and effect

The most effective way to isolate the cause of an out-of-control condition is by drawing on a stock of knowledge acquired by changing conditions when nothing is really wrong. Some plants experiment more than others. Experimenting with production equipment can be costly, sometimes prohibitively costly, so people cannot try something purely for the fun of it. However, struggling through the unknowns when problems do appear can also be very frustrating and expensive.

Before people can experiment effectively, they should be organized for it and prepared to observe and record results. This is part of daily life in development laboratories, but not in factories. Therefore, everyone who progresses to the stage of being allowed to experiment must have developed some expertise and judgment. An objective of maturing operators and supervisors in stockless production is to bring them to a point where they can do a limited amount of experimenting. The more people have done this, the more they know about how processes really work.

QUALITY CIRCLES AND OTHER GROUPS FOR PROCESS IMPROVEMENT

Quality control circles (QCCs) are an entrenched practice in Japanese plants, but have only recently begun to appear in the United States. The QCC consists of 5–12 persons who work together, meeting voluntarily to improve quality or to resolve other problems. Participation is voluntary, but peer pressure to participate in QCCs is higher in Japan. Participation rates there are commonly in the range of 90 percent and above of all employees, while in the United States plants have difficulty bringing participation rates above 30 percent.

QCCs are organized in various ways in both countries. In many the supervisor is the leader of the QCC consisting of those who report to him. Some meet entirely on their own time. Some are paid for one to two hours of meeting time per month. All of them work on projects of their own choosing, but most take cues from the current interest of management in specific problem areas. They do not confine themselves to the quality of the product, but may consider problems of efficiency, layout, safety, or even the personal problems of some of the group (in Japan).

The QCCs build morale because workers contribute to improving their work and their jobs. However, their most important role is in improving the quality of the production process. They typically tackle the small problems necessary to overcome in attaining and preserving process control. They are a very effective mechanism for educating the workers, and much of their training they do themselves.

It takes time and patience for QCCs to become effective. At first they are likely to focus on annoyance problems—for example, how the lines are painted on the parking lot or the color of the trash cans. Then they may take on projects too large in scope—changing the layout of the entire plant, for example. The QCCs need to be carefully nurtured to mature into effective forces for overcoming nitty-gritty problems. To Japanese, *quality* refers to quality of process, not just quality of product, so a broad scope of activity falls under the heading of quality.

In Japan, the QCCs have been essential to the development of stockless production. Without the small improvements they contribute, stockless production would not have succeeded. In plant after plant, the QCCs which had become listless suddenly became revitalized by studying the problems that surfaced when inventory was systematically reduced. As the inventory "water levels" receded, the QCCs were a major tool for "breaking up the rocks" that appeared.

In the United States, it is by no means certain that the QCCs will play the same degree of importance in stockless production. Some Americans are enthusiastic about QCCs; some are not. Plants have introduced stockless production in the United States without them. (Kawasaki USA and American Koyo are two of them.) The Americans did not take to group methods of problem-solving as readily as the Japanese. QCCs have enjoyed some modest success at American companies, but there have also been problems.

Middle managers are not prepared to cope with them. They see the QCCs as usurping their role. American managers emphasize individual responsibility, not group responsibility.

Workers are often suspicious of "voluntary" activity and are unwilling to meet on their own time, and there are arguments over paid time.

Managers had misgivings over the lost time and expense of training, especially if the QCCs appeared not to be producing results of significance while stirring the resentment of middle managers.

The purpose of the QCCs has sometimes not been very clear, so they are considered as having only vague, do-good roles, and make-them-feel-like-they-are-participating projects.

The value of a QCC and some of the issues are best brought out by the story of a QCC at Minolta Camera Company:

> One of the problems taken up by a small group at the Mizuho plant was to reduce the degree of variation in copier lamp voltage at the testing point. This small group consisted of a number of young girls employed on the production line, but as a result of their activities so-

phisticated engineering improvements were introduced into the production process. One of the checks conducted on the copiers, near the end of the assembly line, was measurement of the voltage of the lamp utilized in the copier. One of the problems faced was an unacceptable degree of variation in the lamp voltage of nearly finished copiers reaching the testing station, even though the same units showed a lower variation after undergoing adjustment at an earlier stage of the assembly line. As Exhibit 7–6A suggests, at the adjustment stage, the variation was relatively small, but when units reached the final testing stage, the degree of variation increased significantly. Data shown in the exhibit were prepared by the small group as one of the first stages in their activities to determine the extent of the problem.

EXHIBIT 7–6A Before small-group activities

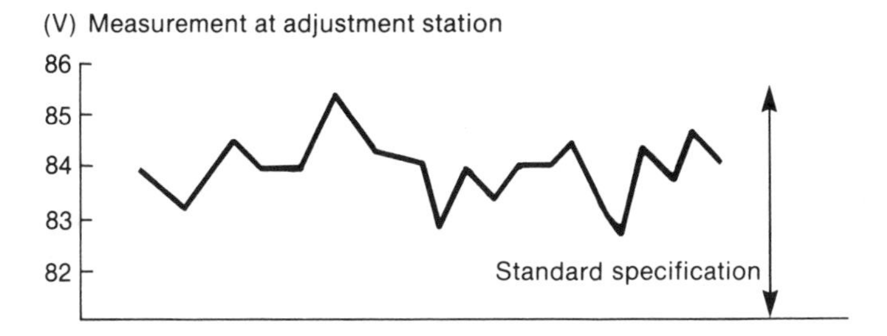

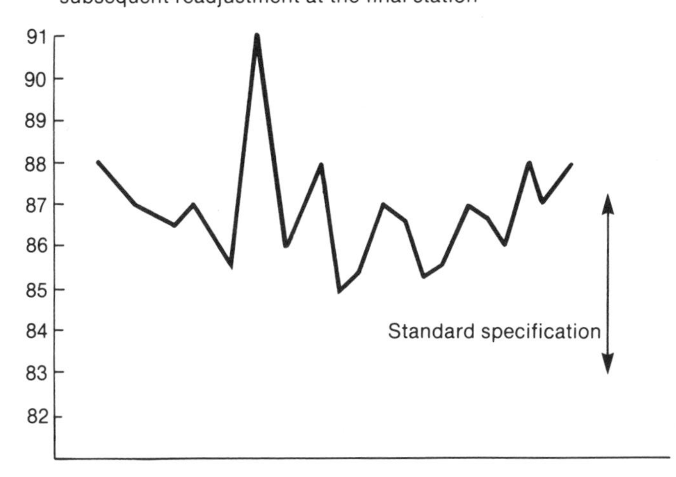

The final results of the June inspection showed that 60 percent of products exceeded set variation standards and that changes of one to five volts occurred between the adjustment and inspection stage.

The next step after collecting data on the variation in lamp voltage was to set a tentative goal for the activities of the group. The group decided to aim at a variation of one volt or less at the testing stage and began its activities of identifying causes and thinking of solutions to identify problems.

The next stage of the group's activity was to identify all possible causes which might increase the variation in the lamp voltage. Various causes which could have affected voltage include methods used by the operator of the testing equipment, the way in which lock paint was applied to the area used in testing in the interim steps between the adjustment and testing points, various environmental factors, and the condition of various related parts. As an initial step, the members of the quality control circle summarized the various possible causes to isolate potential problems. As a result of the identification of possible causes, various means to reduce the error in measurement and other problems causing the variation in lamp voltage were devised. For example, it was found that by applying paint to the point where the voltage was tested between the stage on the assembly line where the voltage was adjusted and the stage where the test was conducted, an error in measurement was introduced. The amount of additional tests and adjustment that this required represented an important increase in cost and a reduction in efficiency. The decision was therefore made to apply paint to the point tested prior to adjustment to eliminate the difference in measurements coming from this source. Another problem was that different individuals

EXHIBIT 7–6B After institution of suggested changes

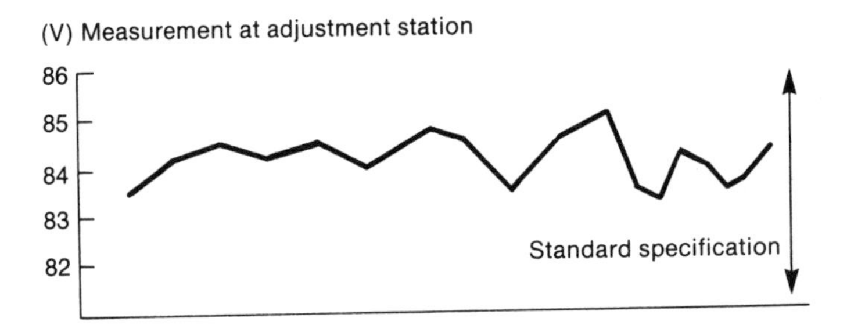

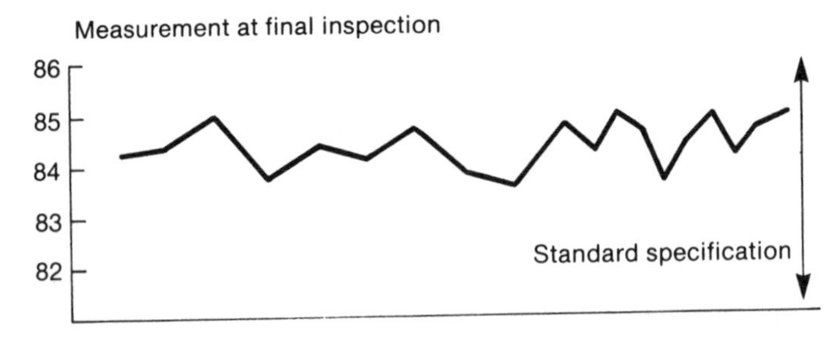

measured the voltage level at slightly different points, leading to errors in measurement. The small group held discussions with all persons involved in measuring voltage and decided on a standard point for measurement.

As a result of better measuring techniques and other suggestions made by the group, the difference in the variation at the voltage adjustment and final testing stage was reduced significantly as Exhibit 7–6B suggests. Comments of the members of the circle in their final report on the project included the following: "We believe we were able to solve this problem because all the members of the circle came together with one purpose and worked to contribute to its solution. The fact that we were able to correct this problem which production process engineers had not been able to solve has given us confidence in our future activities. We want to continue to be active in identifying problems and continue our small group activities.

This story was culled from a publication by JETRO (Japan External Trade Organization). The story is intended to illustrate the power of the quality control circle. Try an experiment with it. Have different people read it and ask for their reactions. Some will say, "Isn't it remarkable for a group of young girls to do something like that!" Others will say, "What is so great about it? It sounds like the kind of thing that happens ordinarily in the course of starting up a line." And there will be reactions in between.

Both of the extreme points of view and many of those in between proceed from the same organizational assumption: Supervisors and staff engineers have the responsibility for start-ups and correcting problems. People on the line may contribute or they may not, but their main contribution is to manipulate the parts. Anything else they contribute is extra.

If a plant is to gain process control and proceed into automation, this role of the operator must change. The operators become more and more controllers of equipment and correcters of deviations. They must contribute through their work to the daily reduction of variance in every form of production activity. To do that, they must be aware of the relationship of all their actions to variance in production activities. No one else can be more aware of the sources of problems than the person doing the work, if only that worker knows the importance of correcting problems and knows how to go about it.

The formal organizational method of the quality control circle may not be the only way to emphasize a new role for operators and other workers in stockless production. The essential goal is to get everyone working every day on all the nagging little problems that prevent process control. At Kawasaki USA, the supervisors play a key role in this. Suggestions are made as individuals, usually to the supervisor, but sometimes to staff engineers. Sometimes a simple improvement is thought of and implemented on the spot with no formality. Some operators are eager to contribute. Others just want to do their job.

Hewlett-Packard Company at Ft. Collins, Colorado, did not have much success with a quality circle program. Working in ongoing groups did not seem to appeal to many of their employees. They achieve a similar result with what they call a corrective action program, a form of suggestion system. The suggestions are placed on a bulletin board, not in a box, and the resulting visibility both induces the suggester to give it careful thought and the supervisor to give it quick attention. The supervisor has 48 hours to either resolve the suggestion himself or have it in the hands of the appropriate person who can set a schedule for studying the matter. The response of management is also placed on the bulletin board.

If the problem addressed cuts across functional or departmental lines, the corrective action request will be handled by a task force, an ad hoc group created for the purpose. Hewlett-Packard has a network of quality improvement teams who participate in this regularly, but some problems may be addressed by a group consisting of an operator, a supervisor, an engineer, and perhaps one or more other specialists as appropriate. When the problem is solved, or the team is stymied, which is not often, the ad hoc group disbands. What this does is allow flexible concentration on production problems within the framework of individual areas of management responsibility traditional to American management.

Hewlett-Packard achieves some of the objectives of the quality circles without creating permanent group relationships where they are not wanted, or creating complications for the responsibilities of individual managers. However, ad hoc groups are not media for education of the workers, as is often true of the quality circles, so Hewlett-Packard has to educate their workers through more formal training programs. That is very important. By making the workers aware of basic quality control methods, methods of work study, and the specific requirements of Hewlett-Packard production, the workforce is able to make very informed suggestions from a firsthand point of view. (Hewlett-Packard at Ft. Collins has a strong quality program in place, but has only started with a few of the elements of stockless production.)

Most important: To make stockless production a success, the role of the worker must expand to that of an accurate observer, conscious of the sources of variance in the production process and what affects it. Then the workforce will become a part of the improvement program. Workers will have an understanding of what needs to be done so that what is desired will be carried out in fact, and they can contribute mightily by observing what is really happening in the absence of supervisors and staff. With good training, they are the means by which stockless production is carried out.

It is attaining this objective which is important, not the exact organizational or cultural methods for doing it. Adopting the form of quality circles without also adopting the objective is likely to be only another passing fad, copying without understanding. If the quality circles are acceptable to the workforce, as they are in Japan, that is excellent because that is an established way of making process improvements at the detail level. If they are not acceptable, other approaches should be tried until an acceptable one is found. As operations shift more toward automation, everyone's job shifts more toward correction and improvement.

8

Product design and stockless production

Excellence in stockless production begins with the design of the product. If it is not designed with production in mind, the stockless production approach can only go so far without bogging down. If design changes frequently, or if engineering changes are not managed well, leveling the schedule and streamlining the flow of material is difficult. In fact, disorganized start-ups and random issuance of engineering changes do much to undermine the system of the shop floor, no matter what it is.

Incorporating new products, new models, and design revisions into production is also a part of the production process, a part by which production is integrated with the other functions of the company. If this process does not go smoothly, production cannot function smoothly. The first few units which come from production are the ones which impress those important first buyers, so it is vital to regard product design changes as something to execute well, not as disruptive exceptions to an ongoing process. Many companies introduce new models and design changes constantly.

As production becomes more automated, this fact must be considered in design if a disastrous level of engineering changes is not to ensue. For example, tolerances for welding by robots are not necessarily the same as for welding by humans. For many products, the tooling and the product must be designed together, and the degree of this integration increases as automation increases. There are both more limitations and more opportunities to designers. The entire area of computer aided design/computer aided manufacturing (CAD/CAM) is directed toward closing the gap between design and production, but the automatic development of manufacturing instructions by computer does not alone ensure producibility.

Design engineering is an activity which should be integrated over a broad scope of considerations, but all too often it is not managed in a very integrative way. Much waste usually comes from design engineers creating a design in a vacuum, then handing it to production to determine how to build it. Some design engineers regard their primary mission as only to incorporate the latest technology into a design. Others are so cautious in making sure that the product works they call for requirements almost impossible to meet by the existing production technology. At both extremes, the designer is not making himself aware of the process control parameters economically attainable in production.

For example, electronics designers may unknowingly select components difficult to handle in production for several reasons: The range of physical sizes or orientations of the components may make it complex to design equipment to move and position them. The components may be difficult to connect or mount. They may not be available in quantity. Sometimes electronics engineers add manufacturing engineers to design teams in order to check whether the evolving designs can be produced by automated means. Consumer electronics companies sometimes locate design departments near purchasing so as to improve communication of the history, and probably future, of circuits selected:

Suppliers may have announced a chip but cannot yet make it to specification in volume.

Suppliers may be ready to discontinue a chip as obsolete.

Suppliers may be having difficulty with the quality of the chip.

Other companies locate design engineers near the marketing department, or they set up design teams as part of a larger team to plan and introduce a new product. The design may need revision, based on tests of customer requirements and reactions.

If a new product or new model introduction is not well managed, the design can be unstable in the early stages of production. Sales personnel may want to promise the product before it can be made, or before testing is finished. If the interaction with the customer's requirements is not stable, any one or a combination of the following can take place:

The product can become overly complex because Marketing wanted too many features in one model or Engineering introduced too many new ideas at once.

Testing reveals some shortcomings, but Engineering is sure it has them corrected with only one round of tests.

Marketing or the customers themselves vacillate on the parameters desired for appearance, price, durability, reliability, and packaging. Marketing can also decide to advance the date when the design should be introduced.

In any case, the model is started into production without major design issues being resolved.

The pressures of competition are such that all these issues cannot be resolved so smoothly that production considerations can be thoroughly treated before a new design goes into production, but it does not help if one or more designers continue at work long after introduction, pouring out a stream of changes to clean up the design. The cleanup is necessary for many reasons besides production: product liability concerns, ease of maintenance of the product, incorporating the design ideas which were almost ready, smoothing awkward configurations, and so forth.

The complexity of the issues surrounding the design of even simple products has increased to where the nature of the design task is affected. Engineering changes begin well before the product arrives on the production floor. Every time a new consideration requires a modification of what is on the drawing board, the opportunity presents itself to overlook something which prevents the design from fitting together as it should.

Few practices lead to more trouble than not checking the producibility of the specifications of a product. If a stack of tolerances is not given a final check, if a product is required to pass an operating test that is on the edge of its design limits, or if the degree of process control economically attainable in production is not reviewed, the seeds are planted for both process control problems and for a blizzard of quality arguments, warranty problems, and engineering changes. These problems in production cannot be solved by asking engineering to be more careful. The practices which lead to them must be attacked. The issues of the rest of this chapter concern mostly the relationship between design engineering and production, but the source of problems goes much deeper than that.

VALUE ENGINEERING

The most common impression of value engineering is that it strives to reduce the cost of a product while retaining the original specifications intact. It may also strive to increase the functional utility of the product while retaining the cost base. In practice in the United States, it ranges from an obsession at a few companies to a subject hardly considered at others. In a stockless production company, value engineering plays a vital role. Ideas to reduce cost may come continuously from production, and many of them affect design in some way. Likewise, any change in design to reduce cost is a change that must be made in production. It is how this process is managed that is important.

The objective of value engineering in the design of the new product is to get as far down the learning curve as possible before the product first arrives in production. In Japan's stockless manufacturing companies, the design teams contain many manufacturing engineers, and value engineering is very important. In this sense, it permeates the philosophy of design.

To make stockless production work well, engineering managers need to understand how value engineering needs to be incorporated into design work, beginning with good design management. Since design is an integrative process, almost everyone needs to have some consciousness of this. Some principles of good design management are:

Have clear goals for the design of a product or model. One of the more successful goals for developing a copying machine was to make it produce copies of the same quality as the present model, but make it half the size and at half the cost. No mush about trying to appeal to three market segments at once, or whatever. Clarity of goals is a general management problem, but they are essential to a focused-value engineering program.

Keep a design from becoming more complex than necessary—simple elegance is desired. This avoids a proliferation of parts.

Complete the design in the appointed time. CAD/CAM is a big help here. The longer the design process continues, the more likely that market changes or technology changes will overtake the original goals of the design, and a process of design turns into a process of never-ending redesign.

The kind of value engineering which reviews parts and production just to niggle a penny here or there may lead to a great many changes without much overall impact, but value engineering built into the design process from the beginning has more benefit. The platitudes above offer good advice, but how to follow them is the problem.

Parts proliferation is an enemy of stockless production. It causes too many setups, and it requires so much inventory that it is hard to organize production to find one part in only one place. Therefore companies have preceded their stockless production with campaigns to reduce the number of parts they use. One company, Tachikawa Spring, achieved a 70 percent reduction in the part numbers fabricated in less than two years. Tachikawa Spring makes seats for vehicles, and these designs go through a regular cycle of design updating. It was perhaps easier for them to do this than for many other companies, but every company needs such a program. Here are several ideas for this:

Create a catalog of internally produced parts, so that engineers can design from the catalog as well as those supplied by vendors. (This usually is not done because the cost is considered too high. However, the purpose is not to create a complete list of everything that could be made, but a catalog of parts that the company would like engineers to use. The catalog can be kept in a computer file.)

Get engineers to review the catalog while the design is still in a formative stage. This prevents getting a design boxed-in so that "dimensions are all about a quarter inch off from what would accommodate existing parts."

Provide engineers with a cost function for parts that favors existing parts. One of the reasons for parts proliferation is that cost functions often consider only the projected cost of a part after everything is in place in production to make that part.

The usual problem in creating the cost function is difficulty identifying from the cost records of a company exactly what costs are pertinent to the existence of more parts. There are many: more parts to carry as spare parts over a period of time, more tooling, more variety of parts to handle in assembly, more setups, larger work-in-process inventories, more records, and more managerial confusion.

This is common sense, but establishing it using the cost system is not so easy. For convenience, usually a standard cost of a part very similar to the one contemplated is used as a surrogate cost. Another variation of the method is to estimate the standard cost from a mathematical function or a cost table developed from the pattern of standard costs created by a family of parts. If nothing more than this is done, the cost comparison totally ignores the overhead costs which come from using a multitude of low-use parts.

The logic of cost comparisons based on current standard costs is that of making incremental decisions, one part at a time. The advantage of having *one* less active part in the system may be almost imperceptible. The advantage of having *many* less active part numbers can be dramatic. Companies that begin to break away from the standard cost comparisons use a comparison of standard cost plus an estimate of the extra tooling. It is very rare to find a company providing its design engineers with a cost function that explicitly approximates the added cost to carry more active part numbers in production.[1]

A better way is to develop a cost function that does not assume incremental changes in the number of parts, but large dramatic ones. Make a study of what would happen if the number of active parts were cut by, say, 30 percent. Then use that study to apply a cost-increase factor to a newly designed unique part. A policy of applying this cost function over time is likely to make the policy effective. At any rate, if reducing the unnecessary proliferations of parts will hasten the time when stockless production can be developed, the benefits are enormous, but continuing with cost comparisons that represent only incremental cost changes by the current methods of production will not lead to large, long-term gains.

If designing to increase the commonality of parts is useful, designing to eliminate parts is even more useful. It is hard to argue that complexity is better than simplicity, but occasionally a cost system can become warped enough to indicate this. It helps to keep in mind what is physically going to happen. If we use fewer parts and they are easy to handle and assemble, and if they are furthermore within the capability of fabrication without defect, the reduced effort means lower cost, no matter what a cost comparison of some isolated phase of this activity may temporarily show. For this reason, a system to guide design by what will physically happen in production is superior. These are developing, and one is described in Exhibit 8–1.

Another reason for both complexity and awkwardness in design is that design teams are too large and sometimes they are too geographically scattered. In this situation, configuration management becomes a problem. In general, *configuration management* means making sure that the overall pieces of a design fit together, something not altogether easy when, for example, the airframe, engine, and control system of an aircraft are designed by different companies in different locations. Methods for coding and accessing the latest in design changes for such projects exist and are used, but, like everything else, they work only as well as the propensity of the engineers to carefully use them.

Most Japanese companies have design teams which are smaller and

---

[1] One company which has developed such a cost function is BDP Division of United Technologies. It is a downward sloping logarithmic function relating cost to number of applications of the part. It was developed by Ilow Roque of BDP, who reported it in "Production-Inventory Systems Economy Using a Component Standardization Factor," *Proceedings of the Midwest AIDS Conference*, 1977, p. 145–46.

EXHIBIT 8–1 UMASS system of design for assembly

This system is the result of several years' work by the University of Massachusetts and the University of Salford Industrial Center in England. It resulted in *Design for Assembly—A Designer's Handbook,* which details factors in easy-to-use checklists which are pertinent to reducing the time and difficulty of assembly. There are basically two types of assembly, manual and automatic, and a different set of tables for each. Start with a basic design of a device and evaluate how it can be simplified for assembly. Following is a summary of the procedure used in evaluating for manual assembly:

1. Estimate the theoretical minimum number of parts required (NM). This is based on the principle that two separate parts in a design can be combined unless one or more of the following conditions apply:
   a. One part must unquestionably move relative to a mating part during operation or service.
   b. The parts must be of different materials for purposes of wear or insulating properties—for example, functional reasons, not cosmetic ones:
   c. Separate parts are necessary in order to permit fabrication of them, or for disassembly during operation or repair.
2. Assuming small parts are to be assembled, estimate:
   a. Minimum handling time (TH).
   b. Minimum insertion time (TI).

   Unless one has better information in a specific case, in most instances one assumes that:

   $$TH = 1 \text{ second}$$
   $$TI = 1 \text{ second}$$

3. Assign degrees of assembly difficulty from tables which relate this to the geometry and surface properties of the part.

   $$DH = \text{Level of handling difficulty}$$
   $$DI = \text{Level of insertion difficulty}$$

The theoretical minimum assembly time for the design (TM) would be the sum of (TH + TI) for the minimum number of parts (NM).

The actual assembly time of the design (TE) would be the sum of (TH + DH) + (TI + DI) for all the parts in the proposed design.

The objective, of course, is to keep generating revisions to the proposed design so that its assembly time comes as close as possible to the theoretical minimum time.

The procedure used for evaluating a design for automatic assembly is very similar. The basic difference is that tables must evaluate the geometry of a part for ease in automated assembly equipment. Are they sufficiently nonsymmetrical to permit orientation by a vibrator, for example?

Provided the company is equipping itself for automated assembly, the objective here is to maximize the number of parts which can be automatically assembled, using equipment which a company has or has planned to acquire.

Commentary on the UMASS system. The UMASS system does not give detailed consideration to the difficulty of fabrication, which is a major source of trouble. In fact, the cost and time of assembly may not be a large percentage of the total manufacturing cost of a product all the way from raw material to finished state.

However, the system begins in assembly, and that is the place to begin. Any part not assembled does not need to be fabricated, which is also a good start toward simplifying fabrication. Likewise, many parts that are easily handled either manually or automatically for assembly are devoid of the complex geometry which creates problems and expense in fabrication. From the viewpoint of fabrication, a serious question is also the size of the part. A single large injection-molded part, for example, may require a more costly die and strain the capability of the extrusion conditions much more than two smaller parts that could be joined together.

EXHIBIT 8–1 (*concluded*)

Natural extensions of this system encompass fabrication. A system to rate parts by fabrication difficulty has a number of problems that may limit its scope to specific industries. Different material selection for the same geometry generally defines the options for fabrication, but as the parts are simplified or altered, the options change. Nonetheless, geometry, size, weight, and material can lead to a rating for several different kinds of common fabrication: pressing, molding, forging, basic machining, and so forth. In fact, the choice of fabrication method is built into computer-aided manufacturing programs that directly develop the instructions for tool-building or machine operation from the part design.

Stockless production companies are known to be moving their design engineering in this direction. The addition of manufacturing engineers to design departments has a more systematic purpose than discussing a design over coffee. What is desired is the codification of the current production process capability and the expected additions to that capability in a form whereby knowledge of it can be built into the details of design development. To be workable, the system has to be easy to use and up-to-date.

The description of the UMASS system of design for assembly was taken from G. Boothroyd "Design for Producibility—The Road to Higher Productivity," *Assembly Engineering,* March 1982, pp. 42–45.

not as scattered as in American companies. This makes integration of the design effort easier, it reduces the problem of configuration control, and it brings closer to reality the simultaneous design of both the product and the process for making it.

American defense contractors use a method of design called *design to cost.* Sometimes their attentiveness to design to cost is perfunctory—intended only to fulfill the expectations of the Department of Defense to the degree necessary, but the philosophy of the method is a good one. It consists of two major deviations from the usual design procedure:

1. The design as originally proposed is assigned a base cost by an estimate. At the conclusion of a design-to-cost activity, the final cost is generally lower than what might have occurred had no cost-reduction effort gone into the design.
2. The cost-reduction program consists of regular meetings and communication among a large number of persons other than the designers. These persons review the design for suggestions on cost-reducing measures. The kinds of persons usually assigned to a design-to-cost team are

   Design engineers.

   Engineering specialists (metallurgists for example, depending on the type of product).

   Value engineer.

   Design-to-cost accountant. (Includes life-cycle cost analysis.)

   Manufacturing engineer.

Quality engineer (quality and reliability).
Purchasing engineer.
Industrial engineer.
Production planner.
Others as required by a particular product.

The idea of design-to-cost may be regarded as a gimmick, but it draws attention to design as an integrative activity. The actions of a design-to-cost team are nothing more than what might be expected if a company is trying to refine its designs as much as possible before they go to production. If integrated design is organizationally awkward, lead times for product introduction are long. *The objective is to incorporate production knowledge in the complete process to reduce total lead time from design concept to smooth production.*

Making engineering design an integrative activity takes more than just a formal program to graft a little integration into a process if all the incentives otherwise seen by engineers point them toward remaining closeted with their creations until they unveil them. The entire process of managing design should promote integration, beginning with educating engineers about why it is important. The design engineer's mission is to bring technological innovation and elegance into products that can meet the variety of criteria necessary to both build them and use them.

As an example of how design engineers have been involved in an integrative activity, Toyo Kogyo in 1974 was in dire straits following the market collapse of the rotary engine, which consumed too much gas for either the Japanese or American market following the Arab oil embargo. Toyo Kogyo took a number of actions to forestall disaster. One was to move into a stockless production program. Another was to send many production people and *design engineers* door to door in Japan selling cars. (Autos are commonly sold that way there.) Any order they took would be helpful, but, at the same time, the *design engineers* got a firsthand education in what the public wanted in cars. Likewise, back in the plants they could see what stockless production was about. From that era, the engineers designed several Mazda models which have been well received since then (the GLC, the RX7 and the 626 in the United States.)

How American engineers would respond to such an assignment is not known, but one can imagine that they would consider selling as not very dignified. It depends on how such projects are presented to them, and on the area in which they specialize. If a technology is moving very rapidly, as in large-scale integrated circuits, the design engineers may have all they can do to keep up technically. However, stockless production is most useful for products which are not on the edge of the state of the art technically, and integrated design is important.

DESIGN OF OPTIONS AND OF THE PRODUCTION PROCESS

Different models and different options are both ways to adapt a single basic product to the uses of many different customers, and at times it may be difficult to distinguish between different option combinations for one model as opposed to the production of different models. Options are difficult to forecast, and they are difficult to incorporate into a level schedule. Large-volume options which have a stable demand are not so difficult, but low-volume options tend to be ordered irregularly, which makes them difficult to produce to a level schedule. Exhibit 8–2 illustrates the problem.

However, it is not true that only uniform products with little option content can be produced by stockless production. It depends on how the options and the production system are planned and designed for each other.

EXHIBIT 8–2 Simplified examples of options in a level model mix

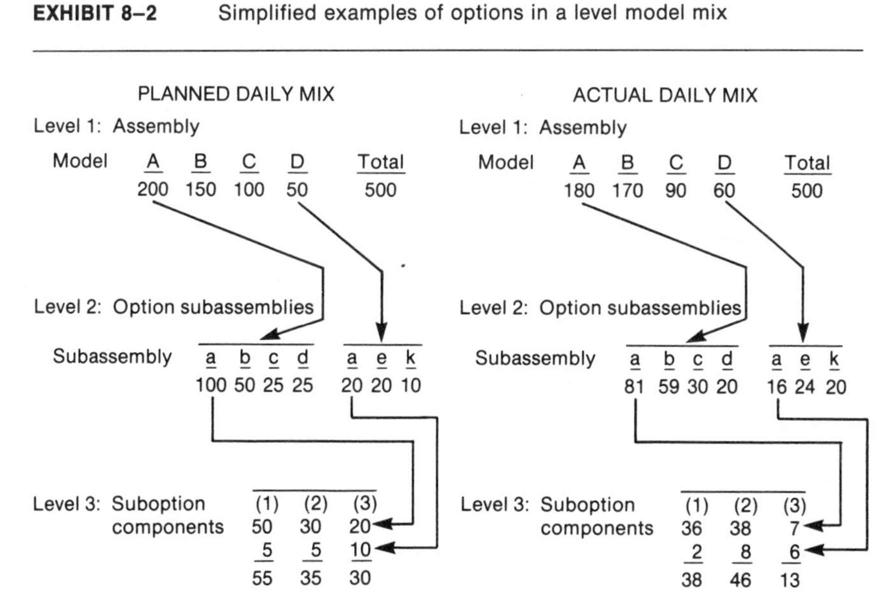

As the complexity of options calls for different components deeper and deeper into the bill of material, the effects on lower-level production demands become profound even when the planned total rate is held constant and only the model-and-option mix changes.

At the level 1 option assembly level, the production rate deviates from plan by a little, and the mix change to which production is asked to respond is very large. Note subassembly k, a low-volume option.

At the level 2 suboption level, the overall rate deviation is more pronounced, and the mix deviation has become drastic, even for the suboptions of the high-volume optional subassembly a.

Toyota Motor Company produces automobiles with many variations. Different countries of destination require right- or left-hand drive, different emission standards, different bumpers, different mirror locations, and different instrumentation. The customers in different countries have different preferences in the final product in addition to the differences prescribed by regulation. However, these boil down to similar complete sets of product differences for each country, what in Chapter 4 we referred to as a *variation.*

A variation can entail a very great difference in the total configuration of the product, but if the mix of variations is worked into a level final assembly schedule, the mix changes seen at lower levels do *not* amplify very much as they work down through the different levels of production in the way shown in Exhibit 8–2. The mix change seen at the lower levels is about the same as the mix change seen at final assembly.

Toyota has historically preferred to offer options of choice to the customer in packages which do not entail big changes in mix in heavy fabrication areas, big press departments, machining, forging, and so forth. Those are the departments that have great difficulty changing from a production plan to react to a major rate or mix change. The goal is to become so flexible in production of some of the lighter production that choices can be offered. Much of this comes down to not offering choices of engine sizes and drive train components other than what is standard for a given model and type.

The ability to react to mix changes is enhanced if a given production department produces all the different components or subassemblies which can come from options and variations. Then that department does *not* see a change in overall rate of production from the changes in mix which occur, and it can work on improving its flexibility to react to mix changes.

All stockless production companies would like to develop the flexibility in every department so that any option could be offered which is within the tooling capability, but progress can only go so fast. In the meantime, one consideration in designing and offering options is the production network as it currently exists. If options must be offered in such a free range of combinations that a commitment to start building components for them must precede a pull signal, we are reduced to building ahead and stocking components at some level of production from which the pull system can draw them to assembly, the traditional approach. Many of them can only be stocked ready for final assembly.

Every stockless production company presently has optional material that fits into this category. In Toyota's case, carpet is a good example. They offer several choices of color and texture. They can only keep the stock as low as possible and draw it as needed. The carpet itself must be made outside the scope of their own production system.

However, the production system needs to be developed to eliminate as much of this as possible. To understand how it might be done, let us review the automotive broadcast system.

Broadcast systems

The actual final assembly schedule for auto plants anywhere is a sequence of specific vehicles with all the detail which are to be assembled on a given assembly line. The broadcast is the sequence of vehicles to be assembled, updated as of the latest moment. As used in the United States, the broadcast provides pull signals for sequencing the material coming into the final assembly line, but it does not always provide for a restriction on the volume of inventory in the pipelines coming in. Its use is generally confined to operations inside the walls of the assembly plant or to immediate deliveries to it, but occasionally it is used to drive major operations feeding final assembly.

Engines, axles, seats, options—everything—are lined up in sequence to meet the correct vehicle at the point where they are mounted. To do this, a broadcast sheet is delivered to all key operations at the start of each day and updated throughout the day. The broadcast is also trimmed in sequence, and it is sometimes given to other operations.

For example, if Oldsmobile sends a signal to the engine plant to begin assembly of an engine in sequence to meet a specific automobile which is scheduled for final assembly, the system is a pull system. The engine plant cannot start assembly until it knows what to build. If the rate of building is constant, then the lag time between when the engine build signal is known until the engine must meet the car provides a system limit on the amount of inventory in the pipeline. If the physical construction of the plants is such that, say, a maximum of 300 engines can be in the pipeline, this fixes an upper limit on the amount of inventory which can be carried in the system. Oldsmobile can vary this amount of inventory to include more or less slack to cover for problems that might occur. If there are no extra engines in the pipeline beyond what are needed for processing, there is no room for error, which means that Oldsmobile expects both the assembly plant and the engine plant to run trouble-free.

On the other hand, the planned line sequence might be transmitted to the engine plant three days ahead of time, and if the engine plant starts to build them three days ahead, the production is really responding to a loose pull signal. The assembly plant is drawing from inventory by a final pull signal if it takes only three hours to assemble an engine.

If the engine assembly is built in either loose or tight synchronization with the actual need for that specific engine on a vehicle, then the exact options are put on that engine in the sequence needed, including the low-volume options. If the volume in the pipeline is kept restricted, then this system can be used to improve operations in the same way as if a card system were used.

In stockless production, inventory can be reduced to levels where fabrication can directly feed final assembly. In fact, this should be done wherever it is possible. For it to be possible, the throughput time of the part or subassembly to be fabricated must be less than the lead time provided by the transmission of the final adjusted sequence for assembly. For example, if the final line sequence is determined at body paint, and the lead time from body paint to the point where a part is to be used is one hour, then any part which is to be fabricated must be producible and ready for use and conveyed to that point on the assembly line in one hour.

Toyota provides a good example of how to handle options under stockless production: Build them in sequence to feed to the final assembly line. Toyota has a project to do this with bumper assemblies entering one of their assembly plants. Up to 100 different bumper assemblies may be required in this plant because different customers and countries of destination have different specifications on bumpers. The project is to give the final assembly sequence to the bumper shop and build in sequence. This gives the bumper shop about three days' lead time. There is some minor juggling of the final assembly sequence during this period, but the inventory of finished assemblies can be kept very small if these changes are communicated. If they are successful in this, they will not have to stock a large number of complete bumpers or bumper subassemblies, but only the basic raw material needed to make bumpers.

Nissan Motor Company, among others, is developing the ability to operate by broadcast systems with suppliers. One of the suppliers is Tachikawa Spring Company, which makes auto seats. Nissan sends Tachikawa Spring a facsimile of the corrected assembly line sequence every 7½ minutes. This is set in final form as soon as bodies have been successfully painted. That is about 120 minutes from the point where the seats must be installed. Tachikawa Spring must within that two hours determine what sequence of seats must be sent to match the vehicle sequence, pull them from their finished goods bank and load them on trucks in assembly sequence to take to Nissan. A truck leaves for the 20-minute trip to Nissan every 15 minutes. There is about 20-minutes slack time in this process. Five to 25 minutes of seats may be on hand at Nissan, depending on the difficulty of traffic conditions.

However, the objective at Tachikawa Spring is to begin the final building of seats in response to the broadcast from Nissan. There are many combinations of colors and fabrics. The further back in the system they can drive the final commitment to build a seat of a given type, the better able they are to handle variety without storing bulky seats.

Another example of this was observed at the Hino Motor Company Truck Plant. Large truck orders specify the wheels and tires desired for the specific application of the truck. The line assembly sequence is transmitted to a tire warehouse a few miles away where the tire supplier selects the correct set of tires and wheels, mounts the tires, and brings them to the assembly plant. There, they are loaded in the line sequence into a chute which drops the correct wheel sets into position for mounting on the truck as they are needed. Flow time from the tire warehouse is about an hour, with deliveries about every half hour; but they receive the line sequence about four hours ahead of time, so normal lead time is about four hours.

Hino also has two engine assembly lines near the final assembly line. The line sequence is transmitted to the engine assembly lines so that the correct options are put on the engines in the engine building sequence in order to meet the truck for which they are intended.

Several feeding operations are synchronized to final assembly by this broadcast system. In effect, Hino is making several subassembly areas function just as if they were part of the final assembly line. The amount of inventory between these areas and final assembly is very small, as small as a half hour of inventory. (Hino does not allow customers to specify engine and drive train options from different manufacturers as do American manufacturers of over-the-road trucks.)

Exhibit 8–3 diagrams a broadcast system used for controlling fabrication and subassembly operations. It may be necessary to remove a unit occasionally from the sequence, usually upon inspection of the body just before or after paint. Therefore, the final correction to the assembly sequence cannot be made until the body clears paint, which appears to be about 2.7 hours ahead of the end of the final assembly line. All the other subassemblies have been committed to sequenced building by then in that diagram, but all the material downline from where the unit is pulled off the line must be repositioned to all for a skip and later reinsertion. Aside from those occasional blips, even the rare options can be positioned in sequence throughout the building process.

If the throughput time of material through its fabrication and subassembly to its point of use can be made short enough, parts can be made in the correct sequence as well as transported in the correct sequence. The more this can be done, the further back in the system the inventory can be pushed, and this can even be done for the parts which are used in a nonuniform, nonrepeating sequence if the production process is capable.

LIBRARY ST. MARY'S COLLEGE

EXHIBIT 8–3 Simplified diagram of vehicular assembly process (subassemblies integrated by broadcast system)

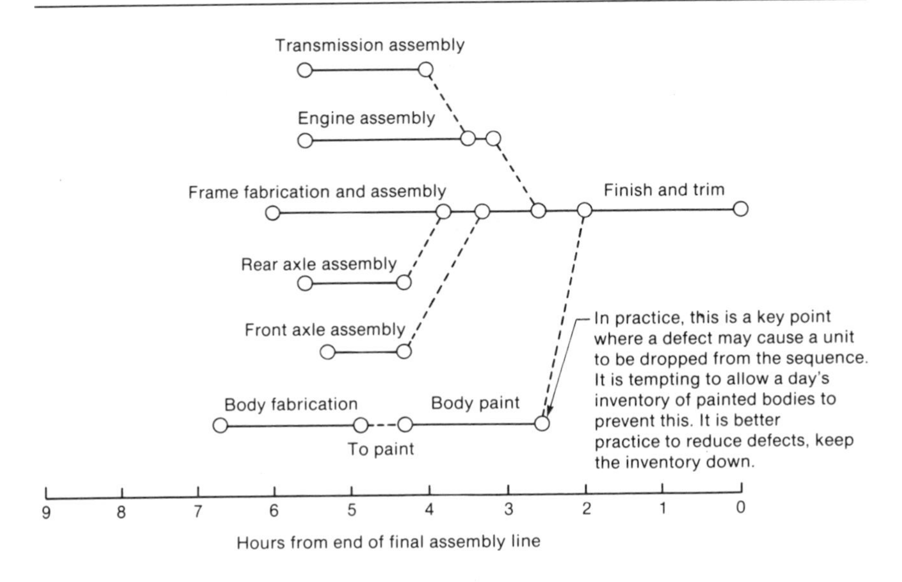

Ideally, the entire process should run the final assembly schedule set for the day. Considering the time needed to sequence parts, tools, or material, that needs to be set eight to nine hours before the vehicle is scheduled to finish. Any changes in the sequence after material is positioned in order must be immediately broadcast to all line and subassembly stations downstream in time. Pulling material out of position and repositioning it later takes excellent communication, or great confusion results. It also takes communication skill to specify repairs on units which have reached the end of the line. Often it is better to make a coordinated stop of all operations and make small corrections, then make a coordinated start.

The Jidosha Kiki Company uses a card system to perform a function somewhat like the broadcast system. For critical parts used on an irregular basis in final assembly, a warning card is sent ahead to the fabricating department for preparation. An example of this is some parts that must be both welded and painted with a throughput time of about two hours. If the parts will be needed at 12, the parts should start into weld before 10. If the warning card comes early in the morning, they know to break their regular routine for the expected quantity on the warning card. The actual move cards would arrive a little before 12. This prevents keeping a supply of optional parts in an outbound stockpoint "just in case."

What is required to produce a large variety of optional items by a pull system

The production process must be extremely flexible to handle the large variety of volumes and mix assortments that come from many options. If an option which is rarely used can be made and transported to the assembly point when needed by a pull signal, that would prevent the usual waste of overruns and obsolescence. Sometimes this is possible to do with current technology, but much more remains to be done to develop stockless production to do it well.

Much more has to be considered than whether we can develop a final assembly sequence to be operated in cadence. In all the examples discussed in the preceding section, the production of irregular options created a challenge to make a few steps of production closely synchronized to final assembly without destroying the advantages of balance and rhythm in production. This obviously is easier to do if the option is an easily fabricated add-on that does not require a large number of coordinated changes.

Summarizing what is required:

1. Ability to execute rapid setup, which depends on
   a. Compatibility of the option's tooling requirement with the physical procedure for quick setup.
   b. Ability to rapidly select the correct kit of items for setup when requirement for it may give little or no forewarning.
   c. Ability to physically and financially store the required tooling (and perhaps material) so that it does not disrupt the flow of production. (If a complete setup were left intact, there would be very low utilization. The same is true of tooling even if rapid setup is possible.)
2. Ability to convey the option to the point of assembly quickly:
   a. Does space allow fabrication close to final assembly?
   b. If not, how much lead time is required to fabricate and convey it there?
3. Ability to call for the option in assembly and install it while holding production linear, preserving balance in the system.
4. Ability to do all this while preserving the ability to perform the necessary quality checks on the option and preserve the quality system throughout the rest of the production network.

Consider as an example one of the more vexing problems associated with a large number of models and options, the differentiated decals and literature for each one. Suppose it were possible to make these things on the assembly line by a copier immediately preceding the time they are attached. For some of the simpler cases, this is within the realm of current technology. The master copies can be kept in a quick-access storage, and reproduction could be done quickly, and it may even be possible to cut them to different shapes quickly. However, the space required, the handling, and the organization of the task must all be extremely simple and failsafe in order to be practical.

Computerized technology makes it possible to sort, organize, and recall options if attention is given to it, but a physical sort is something else. A sort of the masters for decals and literature may be entirely electronic, and is in a completely different realm from sorting through two-ton die sets. Size is an important consideration. Tokai Rika is able to store, sort, and select a large sequence of small die sets for making different cuts on keys for auto ignitions. The press used for this is the one which is capable of exchanging dies with every stroke of the press if necessary. It seems feasible to make a large variety of key cuts by other means, for example, by calling a sequence of programs for a computerized grinder.

In fact, the methods for selecting and organizing options by computer can be very complex. Schlage Lock Company uses design analysis tables for this rather than doing a conventional sort of preset options. A "scratch pad" is defined in their computer program for each separate lock order received by Schlage. Any order may be different from any other order received in the history of the company, and the total number of component combinations possible is in the trillions. The scratch pad records the components selected for the order by referring to design analysis tables that denote different conditions of use for which they are necessary. These conditions of use vary according to door thickness, right- or left-hand door, knob type, lock chassis requirements, and so on. The incoming orders specify the conditions of use for the order, and the design analysis tables convert these to a selection of specific components by an associative logic. This overcomes the problem of having to sort through an enormous memory bank of possible option combinations, which eliminates the need to maintain an enormous updated memory bank as well as greatly reduces computer processing time and increases order definition accuracy.

However, Schlage is not a stockless production company. They schedule the components into an inventory from which final assembly can draw them. Were they to try to convert to a stockless production mode, they would try to cut fabrication throughput times until final assembly could draw them directly through the production system. Leveling the final assembly schedule to balance all production for this might be possible, but it will take intensive thought on how to do it—beyond the current level of practice anywhere.

There is much left to do to further develop the capabilities of stockless production by those who are involved in it, and most of that development will go into how to handle a larger variety of options by a pull system. Much more can be done to integrate all the potential technology into a total production system that is both simple and does what is really wanted. However, development of stockless production starts simply, with what exists, so the engineering of options should take into account the current state of development of the system.

Design of options for production by pull systems

When done for a stockless production system, all design, but especially that for options, needs to be done with a more thorough review of the production process than is usually the case. This would ideally be done by design selection using process tables somewhat like those discussed earlier (Exhibit 8–1) for value engineering. It may be possible for a company's manufacturing engineers to make up a systematic approach considering the current state of stockless production in a given company. An overview of the things to consider is in Exhibit 8–4, but in a very general form. A similar set of considerations applies whether an option is to be bought or made, but, of course, if it is made, specific knowledge of the supplier's system would likely be more limited.

In summary, the most immediate objective in determining how to handle options with stockless production is to simplify them to a point where a level final assembly schedule can be planned. Balanced production will work if the number of irregular options is small enough that the company does not have to revert to operating production through a stockroom system.

NEW MODELS AND ENGINEERING CHANGES[2]

Most important is that the product be designed to produce from the time it is in concept stage. If Design Engineering turns the product over to Production without this, it takes time for familiarization in production, and it takes redesign to make complex products run in production. Even if the product has been designed for producibility, there are still oversights, so every company must put energy into the interface between design and production.

Stockless production companies have a variety of methods for starting a new model or new product, but all of them emphasize a lot of small producibility changes as soon as the model starts. The objective is to work through as many of the small revisions as quickly as possible so as to establish a stable bill of materials for production. It is not greatly different from start-ups at any manufacturer except for the intensity of revisions on the spot during start-up.

[2] See also the section on changeover between schedules in Chapter 10.

EXHIBIT 8–4 Design of options for production by pull systems

Usual considerations for producing ahead of final assembly	*Usual considerations for producing by pull signals from final assembly*
Fabrication: 1. Equipment and tooling feasibility and cost. 2. Review of total fabrication levels for labor and overhead. 3. Material cost and material yield.	Fabrication: 1. Equipment and tooling feasibility and cost. 2. Levels of fabrication and subassembly. 3. Ability to quickly set up at each level. 4. Effect of potentially nonlevel requirements on mix changes at each level. 5. Effects of mix changes on balance of operations at various levels (the fewer the better). 6. Sizes of pipeline stocks, if needed. 7. Material cost and material yields. All of the above says a lot about labor and overhead costs, in physical effort, if not dollars.
Assembly: 1. Configuration interference; "assembleability." 2. Operational feasibility: *a.* Ability to sort and handle. *b.* Line balance. *c.* Stocking levels.	Assembly: 1. Configuration interference; "assembleability." 2. Ability to work into a level final assembly schedule. 3. Operational feasibility: *a.* Ability to select correct option and correct combination of options at the required time. *b.* Line balance; effect on linear production. *c.* Added space for fabrication close to its assembly point at the required time.

This is one of the most complex aspects of stockless production, so this is only a starter list.

All manufacturers start with a prototype which is reviewed for production purposes to plan the start-up. Of course, some of the tooling may already be on order at this point.

Products that are not overly complex usually have a trial run. For example, motorcycles and appliances are products for which 5–10 units are assembled on a trial basis. This trial may take place in a slack period at the end of a shift or between two different planned schedule periods. The purpose of the trial run is to assure that parts can be made with the production process in control, that all the parts are accounted for in planning production, and that all operators have instructions for fabrication and assembly as needed. If the trial run shows that everything is in readiness, the model is added to the schedule for production in an upcoming planned schedule period.

For more complex start-ups, the process takes place with production slowed to a crawl while everyone checks for problems. Many engineering changes are done on the spot. For example, Toyota says that they have 3,000–4,000 running changes during the start-up of a new model. These are accomplished by working closely with the production floor over the first weeks of start-up. Only after the bill of material and production instructions have stabilized do they go back into squeezing the inventory levels and resolving normal production problems.

In either case, the design engineers, manufacturing engineers, production control personnel, quality engineers, and everyone else who is necessary live on the production floor during start-up. Engineering changes are made on the spot. Many of them are small ones: A hose cannot follow the path it was intended to follow. A different type of fastener must be used. Holes do not align as planned. A direct engineering assault to correct these in the beginning prevents the need to dribble a constant stream of engineering changes through the formal system over a long period of time.

At the time the new product or model starts into production, the bill of materials is given to Production Control for maintenance. That department has a very strong interest in seeing that the bill of material correctly describes how the product is really made, but Engineering does not. If a great many changes are made at start-up, the bill of materials must reflect the changes without delay.

In most American manufacturing companies, the engineering department keeps the bill of material. They may be reluctant to give up the practice for fear that someone other than qualified engineers will make a serious blunder in changing the design or material used. This is a valid concern, and the engineers should retain the opportunity both to make design improvements and to reject those proposed by others that risk a compromise in the design of the product. Most companies have engineering change procedures which require that Engineering put its approval on design changes. That is not the same thing as having a bill of material that accurately describes how the product is really being built.

The responsibility of Engineering is to see that the product as it is built does what was intended in the design. If it does this by making and approving changes when required, the mission is accomplished. If Engineering sends out a flood of engineering changes without much regard for the consequences in production, they may be making their own life more difficult than it should be as well as contributing to inefficient production.

Two main engineering actions are necessary to make production flow smoothly:

1. Work through as many engineering changes as possible when a new model starts into production so that, at the close of start-up, it is ready to run smoothly.
2. Thereafter, group engineering changes to take place at the times when there is a major change in schedule.

If either of these is not done well, the result is a stream of engineering changes into production. A pull system which is in operation cannot accept engineering changes except by going to the plant floor and interrupting the process to install the change manually. (This really is true even if the plant runs by a scheduled push system. The change can be inserted in the schedule, but it is executed by making a manual intervention in support of the schedule. *Someone* has to remove the old parts and start the new ones.)

If the plant uses a pull system with cards, Production Control must go into the plant and intercept the obsolete parts and start phasing them out a few days before the change is due in final assembly. The substitutions or changes are then walked through the work centers through which the parts must pass on the way to final assembly.

Of course, every company occasionally has to make an engineering change for a reason so urgent that the efficiency of production must be disregarded, but most of the engineering changes are not so vital. Stockless production companies generally plan engineering changes one to two months before the effective date. The effective dates of most changes are timed to occur at schedule-change times.

The calculation of least-cost run-outs of old parts is no different with stockless production than with any other system. However, if only a few days' of material are on hand in the plant, this task is not hard. In stockless production, more attention has to be given to the effect the change will have on the balance of the material flow through the process.

Some engineering changes take place almost without the workers knowing about them. For example, if a slightly different screw is to be used, a box of different screws can be taken out to the floor for the workers' use and some of them may never know.

A highly complex engineering change is more likely to require the kind of attention given to model start-ups. New instructions, rebalancing, and different material flow paths require explanation and possibly some practice.

EFFECT OF STOCKLESS PRODUCTION ON DESIGN ENGINEERING

The primary effect on Design Engineering is to integrate design closely with other functions and departments of the company. Integration with production has been emphasized in this chapter, but that cannot be done well if Design Engineering is not also well integrated with Marketing and other considerations vital to presenting new designs in finished form to the customer. It is very difficult for the design engineer to be fully aware of all the considerations which should influence the design of a complex product, so this provides a strong impetus toward team design, or at least toward reviews of design by multidisciplinary teams. This is not a new subject for engineering. Many corporate engineering departments and many engineering schools have seen that a multidisciplinary approach to engineering will be the way of the future. Stockless production brings that future a little closer to the present.

9

The supplier network

The supplier network of Japan draws immediate attention as an aspect of stockless production by Americans, and one of the misconceptions is that stockless production is primarily a program to deliver materials from suppliers to industrial customers in small quantities and at frequent intervals. This is further seen as basically a program of inventory reduction by this means. American suppliers of companies that announce such programs fear that the real intent of their customers is only to shift the burden of carrying their inventory to them without compensation.

However, as we have seen earlier, stockless production has much larger goals than just inventory reduction. The intent is to strengthen the production capabilities of both customers and suppliers. Inventory may temporarily be shifted to suppliers in the course of doing this, but the long-term goal is to reduce inventory everywhere; and as the inventory is reduced, it soon becomes less a measure of successful reduction of working capital than a standard for measuring the production process capability of everyone in the production network.

The history of suppliers and their industrial customers in Japan is very different from that in the United States. Japan has provided the environment in which the idea of a physical supplier network has begun to bloom into a reality, but the tightly-linked supplier network is possible in many countries despite a difference of geography and culture. The evolution of supplier linkages in the United States must follow a different path because of the differences in cultural and legal traditions.

HISTORY OF SUPPLIERS TO JAPANESE MAJOR INDUSTRIES

Fifty or 60 years ago, major Japanese original equipment manufacturers (OEMs) had very few companies on which they could call to provide a reliable source of material and components. This was true in many fields: auto, electronics, photographic equipment, appliances, and others. Therefore these Japanese OEMs had to do much work to develop their own base of suppliers over a number of years. This has led to a number of practices which seem strange in the United States.

Consider the reasons why companies have suppliers, omitting the obvious one that no company by itself has the financial resources to do everything for itself from raw material to finished product:

Type I: The supplier can deliver materials or parts at a lower cost than would be true if the customer tried to do it. These suppliers may have lower labor costs, or they may be able to use more specialized equipment because they supply similar material to many customers. In so doing they develop a certain amount of expertise, but that is not the primary reason customers buy from them.

Type II: The supplier has skills not possessed by the customer and not easily developed by them. Because of technical knowledge, the customers call on them to help with more than just providing a basic product. Customers call on representatives of steel, paint, and electronics companies to assist with problems they cannot solve themselves. Some supplier companies start with patented ideas, which form the basis of the product and service they offer to industrial customers.

Because of its industrial history, Japanese suppliers frequently started more as type I suppliers than as type II. Many more American suppliers started as type II. Japanese OEMs frequently started supplier companies themselves in order to obtain lower costs. Sometimes a piece of equipment could be moved from the customer factory to a small shop where it could be operated at lower labor cost. Sometimes retired executives of the OEMs started such companies, thus providing a service both to themselves and to their former companies.

The benefits of this practice were so obvious that the OEM often loaned money to entrepreneurs who wanted to start supplier shops. If the supplier was financially weak, the OEM might allow them to work on inventory on consignment, and the small shop continued to have access to the engineering and production talent of the mother customer. All the large Japanese OEMs continue to have sole-source suppliers of this type, frequently called subcontractors. Most subcontracting in the United States is limited to operations such as plating and coating, with exceptions in such industries as aerospace.

Japanese think of attaining efficiency through close supplier relationships, maintained through thick and thin. All the OEMs and many of the large suppliers have large families of suppliers, some of whom supply the mother company exclusively. Americans think of attaining efficiency by all potential suppliers having a "fair shot" at the business. Most suppliers are much more independent of their customers and proud of it, and both they and their customers prefer a degree of arm's-length negotiation in their relationship. The legal frameworks of both countries support each of these two ways of thinking, which is one reason why trade negotiations sometimes degenerate into confusion. One does not become a major supplier of a Japanese company without becoming at least a distant relative, if not a close member of the supplier family. Trust and loyalty are everything in that system.

The Japanese supplier arrangements are now very mature, and they have become complex enough that it is hard to cite specific practices which are typical of all supplier relationships. The Japanese themselves have several different names for several of them. The Shitauke are typically small and closely tied to the customer company. The parent company typically buys all or most of the output, and they set the specifications of parts to be made, so the Shitauke are usually small type I companies that exist mostly to provide low cost and to improve the focus of the customer company's plants.

The Gaichu are more distant relatives, but still strong members of the family. During this phase of development the supplier company begins to supply other companies and develop technically.

The Kaubai are almost independent. Steel companies are an example. These companies are more financially and technically advanced, or separate; they supply the commodities or more technical components. Their role is more that of a type II supplier.

However, the traditions of the small company dependent on the customer permeates the relationships of almost all industrial suppliers and their Japanese customers. Sometimes even the type II suppliers lay a detailed cost estimate on the table when negotiating price. Both companies have an interest in looking at the details of manufacture in order to improve quality and reduce cost. The price to the supplier may be thought of as consisting of two parts: one to cover the cost of making the components provided, the other a fee reflecting the kind and quality of service that the supplier provides. There is more to being a good supplier than just providing the parts FOB on the supplier's shipping dock.

Occasionally the payments to a small type I supplier reflects the system overtly. The payment system assumes the supplier works on consignment:

Processing fee (per piece × number of pieces)	$100
+ Adjustment for material variance	−50
	$ 50

If the use of material deviates from what was planned, for any reason (scrap, misplacement, damage, record accuracy, or whatever), the loss comes out of the supplier's fee. Under such a system, the customer is immediately informed if there is a problem.

Payment systems may allow for penalty payments of various kinds—perhaps a payment for stopping a customer's line if material is not delivered in time. Often the arrangements are not very detailed when agreed to, and a penalty payment which could be extracted is not actually demanded, but the implication is there. It provides a setting in which very big obligations are felt and in which there is a minimum of arguments over whose fault it was when problems arise.

As Japanese companies grew, many of them outgrew the dependency that existed when they were only a small shop with one customer. They became financially independent. They began to carry their own inventory. They developed technical expertise. They began to form supplier families themselves. However, the tradition has continued, so that even the relationship of large and strong suppliers with their industrial customers is never as independent as in the United States.

This is evident in the way Japanese companies negotiate with their customers. "type II" Japanese customers may not lay a detailed cost estimate on the negotiating table, but they are expected to be very open in recommendations for improving the cost and quality of parts and materials supplied. The supplier company has good assurance that information divulged to customers will not be used to take business away from the supplier. (That is only done in cases of drastic nonperformance, or if economic conditions become so bad that there is no other choice.) The relationship is one of great mutual assistance, so that suppliers are expected to assist customers with their problems in any way possible. Therefore an industrial OEM may give a type II supplier only the minimum critical specifications on a part to be supplied, accompanied by the unspoken expectation that the suppliers will provide the most appropriate material and detailed design to make the part work well in production and in service. (American companies supplying Japanese customers may be disconcerted by this practice.)

Supplier relationships are long term. Bidding on business may or may not be used, but it is used much less than in the United States. Unique parts are negotiated, not bid. Parts which might be purchased by bid might be electronic circuits, for example, for which product life cycles are short and learning curves are steep.

Fostering personal relationships at several levels is important between Japanese companies linked in this way. Therefore, the job of a Japanese purchasing manager is very different from that of his or her American counterpart. Retired executives from customer companies sometimes take positions at suppliers. The purchasing manager may deal with people he or she knows well, seeking to make all personal contacts between people at the two companies beneficial ones. The position is more one of facilitator and coordinator than of negotiator.

In many American companies, the purchasing agent must do many things to improve the cooperation between the two companies. Quality and delivery problems must be attended to, for instance. American purchasing agents become more involved with operational detail than they often would like. They picture themselves as accomplishing more through negotiation than is often the case. The same tasks are accomplished, but usually more by direct action of the purchasing gent than by facilitating the work of others. An American purchasing agent who insists on being the only point of contact between customer and supplier can become a disaster, but the practice has its grounds in experience. A supplier who gets more than one set of specifications from two or more engineers at the customer company is confused. A contract with an American supplier can go sour for many reasons, and the purchasing manager must constantly monitor the performance of suppliers with alternatives in mind. He does not have the tradition of long-term relationships to fall back on. In fact, a purchasing agent pursuing the kind of working arrangements used by Japanese companies could run afoul of a number of laws.

Japanese companies developing stockless production cannot assume that their suppliers will improve internal operations on their own. All major Japanese companies have supplier associations. If they start stockless production, they use these associations to hold regular meetings of many personnel from supplier companies to explain what is taking place and why. Very few American companies have any comparable method of communicating in detail to suppliers. These associations hold technical conferences, and sometimes they form special committees to deal with specific problems.

One of the most famous of these supplier associations is the Toyota Supplier Association. Its major activities are:

Annual and semiannual conferences on industrial engineering and on quality control.

Executive conferences to announce and discuss the specifics of new products and models.

Meetings to discuss system changes.

In addition, suppliers and customers often exchange visits. These are done regularly, and technical personnel not only visit each other's offices, but also their plants. When a supplier company wants to start stockless production, they can often send prospective implementers to work for a while in a customer company's plant or go to a supplier company in the same family to learn about it firsthand. Toyota has had hundreds of prospective stockless production project leaders come to Toyota plants for on-the-job training. The supplier networks are set up for effective people flow as well as effective material flow.

In this kind of supplier network, it is regarded as much more important for people to maintain confidence in each other than to shift allegiances because of temporary and minor cost disadvantages versus other suppliers. In fact, the customer regards it as a duty to assist the supplier in getting costs down if this happens. The real market competition is for the buyer of the end product. The more quality at lower cost, the more everyone in the chain of supply will benefit, so the emphasis is on improving the entire production process from raw material to final product.

JAPANESE SUPPLIER TRANSPORT NETWORK

Because of the long-term nature of supplier relationships, a supplier is regarded for purposes of material acquisition as another work center, which happens to be outside the walls of the customer companies' plants. The management of suppliers is based on maximizing the effectiveness of the total network of suppliers. People in the system cooperate to maintain process control and reliable delivery systems; they concentrate on the production process, not on differences of a few cents per part.

Over time, this Japanese development has led to a network of suppliers providing a large percentage of material to each Japanese OEM through a more narrow network of suppliers than is true in the United States. According to Japanese experts, the typical Japanese manufacturer has about 70 percent of its cost of goods sold represented in purchases, whereas an American manufacturer more typically has about 50 percent of the cost of goods sold represented by purchases. All of this comes through fewer suppliers in Japan. For example, Toyota has about 210 suppliers of automotive material. Ford Motor Company has about 7,800. The Japanese supplier network is typically much narrower than an American supplier network, and sometimes a bit longer. The American network is bigger, and more of a jumble of loose or occasional relationships, as depicted in Exhibit 9–1.

EXHIBIT 9–1

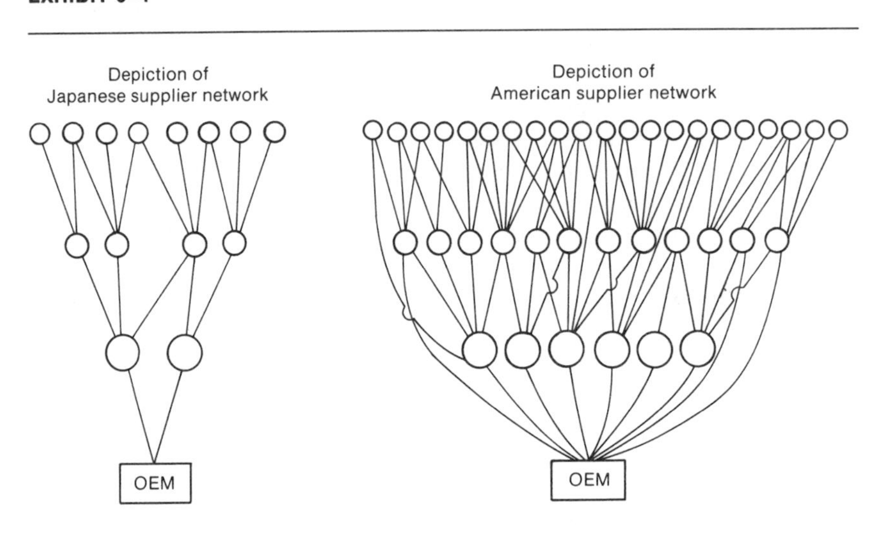

The evidence clearly shows how the supply networks of Japanese manufacturers has developed like a narrow funnel into them. There are many instances in which one part is found only in one place. Sole-source suppliers are very common, and this promotes tight material flows between companies. Supplier companies often build a plant near one of their customers when that customer builds a plant in a new location. The historical prevalence of type I suppliers has promoted this, and it fits well into the pattern of stockless production. Points of supply are geographically close, the amount of pipeline inventory is small, there are not strikes of any duration, and there seems to be little fear of the impact if a fire in a single supplier plant wiped out a sole source of supply. The results of this in several companies is shown in Exhibit 9–2.

Another misconception about stockless production is that companies merely push their inventory onto their suppliers. That is not true of the large suppliers, as evidenced by the numbers in Exhibit 9–2. It is not true of the small suppliers either. On a recent trip to Japan, two American visitors asked to see small suppliers at the tail end of the supply chain and walked in unannounced. They visited two plants, each of which had less than 25 employees, and both of which were working on consignment. Neither company had more than two days of stock on hand.

The point of cutting inventory in the pipeline is to cut the time between the supplier's last operation and the customer's first operation at which the same material is used. *This pressures both supplier and customer to improve the quality of the production operations everywhere—the real objective.* Although the supply pipelines of the United States are much longer and would therefore contain more inventory, the principle and the objective are the same—improvement of the operations themselves.

For a smooth flow of material, the ideal transport system is the one which comes closest (at practical cost) to the effect of handing one part at a time from the last operation of the supplier to the first operation of the customer. Japanese transport systems have evolved closer to that ideal than has the system of any other industrial country. Japan is so small that a supplier cannot be very far distant and still be in Japan, but the supplier transport system did not evolve from the circumstances alone. It resulted from plant location and transport policies which have been employed over a period of years:

Locate supply plants close to customers. Supplier companies build plants close to major customers. Location as part of a major supply network predominates over other location factors (such as the desire to avoid unions might be in the United States).

Use small, side-loaded trucks (22,000 pounds is the maximum legal truck weight). The rail network was not developed to handle a large volume of freight, and the roads are too narrow to use trucks the size of American trucks. Since the trucks load from the side, truck docks are unnecessary as with large American end-loaded trucks. A fork truck can unload one skid from the side and load another skid without the truck being specially positioned.

Ship mixed loads.

—The side-loaded trucks travel fixed routes of five or six stops daily according to a fixed schedule, like a bus route. Some routes are traveled daily; some every other day. At each stop the truck can take on one or two skids and off-load one or two skids.

—Ship mixed loads on both small trucks and on larger vehicles, then there are no economics to encourage full truck loads on only one item. This is extended to rail and ship transport also. (Koyo Bearing ships a mixed load of all materials on one ship a week to its American plant. That cuts the amount of material needed at both ends of the line to only about one week's worth. Kawasaki ships a mixed cargo from Japan to Nebraska with daily arrivals of a mixed cargo ready to take to the line.)

Use load-switching points. Warehouses may serve to hold inventory, but they often serve only as switching points for trucks to exchange loads. In that way, trucks traveling fixed routes can

EXHIBIT 9–2 Size of Japanese supplier networks and stockless production delivery systems

Company	*Number of first-tier suppliers for parts and manufacturing material*	*Supplier distance**	*Transport time**	*Delivery frequency*	*Raw material inventory*
Toyota Motor Company	Approximately 300	90% within 300 km	4–5 hours	• Most parts daily • Engines, transmissions,‡ bulky items: every 15–30 minutes	?
Nissan Zama Plant (auto assembly)	Approximately 130	?	?	• Most parts daily • Engines and transmissions:‡ 12 times per day • 5 parts: every 15–20 minutes	Under 1 day for all major items
Hino Motors (integrated heavy truck)	Approximately 300	90% within 100 km	2½ hours	• 85% at least daily • Engines: 5 times per day • Wheel sets: every hour	Under 1 day for most items

Jidosha Kiki Company (brakes and steering: supplies Toyota, Hino)	Approximately 200	?	?	• Over 50% daily • A few 3 times per day	1.0-day average
Tachikawa Spring Company (vehicle seats: supplies Nissan)	120	?	?	• Most items daily • Bulky items twice daily	1.0-day average
Tokai Rika Company (dashboard instruments: supplies Toyota)	162	?	?	• Alloys and steel twice a week • Most parts daily • A few parts 4 times per day	2.5 days for raw material plus purchased parts
Hitachi Sawa Works† (auto plus other parts: supplies Nissan)	Approximately 400	90% within 50 km	1½ hours	• Alloys and steel, 2 per week • Other: 10% every other day 20% once per day 50% twice per day 20% 4 times per day	?
Kawasaki Motorcycle Division	?	90% within 300 km	5 hours	?	About 1 day

* Transport time and distance are both important. If one is trying to reduce pipeline inventory, frequency of delivery and transport time determine that. When asked, most of the executives could not give the average transport time and distance with accuracy, but they reckoned that the majority of their suppliers transported less than 100 kilometers and had a transport time of two to three hours on the average. The transport times are not easily estimated from distance because of the practice of sending trucks on a circuit of plants and warehouses.

† The Hitachi Sawa Works is very large, with a number of plants within the works making a large variety of product. Some of them operate as job shops, not as repetitive manufacturing.

‡ Auto or truck assembly plants typically have some large items arriving very frequently in line sequence, for example, car seats.

move loads between points on a combination of routes. They also serve as consolidation warehouses for clusters of suppliers.

Use small, nearby warehouses for "valve" inventory. Some suppliers may be outside the transport network covered by the daily truck runs, and a few are even in foreign countries. They may ship in truckload and larger quantities. The using plant will obtain material from the valve warehouse by the card system. The plant personnel do not want it taking space in the plant or being used as a crutch to avoid facing process problems. (The supplier owns the inventory until it enters the plant where it is used.) Of course, not all suppliers may be developing stockless production, so they produce in long runs, and they may also ship to the valve warehouses by the same system. However, they will probably store their own inventory, if possible, to avoid double handling.

Almost everything except heavy bulk commodities moves by truck, but there are no truck docks in Japanese plants. In assembly plants, the trucks may deliver straight to a point along the assembly line, but in most other plants the trucks deliver to receiving departments. These are usually small and serve only as areas for collecting and inspecting. Since there are no docks, the areas are also small, sometimes no bigger than an American two-car garage with a door big enough for a fork truck. Much loading is done outside, so a canopy or other protective cover may be provided over the unloading area. (The weather is seldom severe.) Occasionally, trucks will deliver directly to a point of use in fabrication other than a receiving area. There is no sense moving stock twice unless the imperfections of the system require it.

Workers in receiving check the quantity of incoming material. They use some method for recording a running total by part number, often by an optical scan or a bar-code scan of the cards. Sometimes a summary card is used, to avoid handling a volume of cards. The running cumulative total of each part is used to add to inventory counts and to develop monthly invoices for the suppliers. Usually suppliers are paid for what they have delivered. For those suppliers on consignment who do not use the card system, tracking the inventory in their possession is the most nettlesome inventory problem.

Shipping rooms are likewise small and similarly designed, and they are often the same as the receiving area. If the plant is shipping to customers who are themselves using a card system, the shipping room is the key control point for the cards received from those customers and therefore a key station for the production control department to transmit the customers' cards into pull signals for the plant.

The same truck drivers usually drive the same route each day, so they know the stops and the cargo well. They may regard themselves as material handlers as well as truck drivers and help load or unload the trucks. Sometimes the customer and suppliers jointly own the trucking company, so that, in effect, the truck drivers are also part of the supplier family. Since they often help with the material check-in procedures because they are responsible for carrying the pull signals as well as the cargo, it is very important that they keep that part of their function well-organized.

In order for this system to work, a great deal of attention must be given to the trucking schedules. Truck arrivals and departures must not be bunched too greatly into any one time during each day, or the material could not be processed in and out smoothly and the trucks themselves would be delayed. If pipeline inventory contains only a few hours' inefficiency stock, there is great emphasis on adhering to schedule.

This method of transport has the advantage of promoting careful handling. The less handling the better, and the material is usually handled by persons accustomed to it. Special racks and containers are more likely to be treated with care and accountability.

Security seems to never be a problem. When inventory is limited and all of it is visible, it becomes more difficult for material to just disappear. Tracing a shipment is virtually unknown. Of course, a determined person can foil any system, but the real security in this one is that unauthorized withdrawal of anything at any point is likely to cause a problem noticeable to several people.

The objective of this system is to come as close as possible to a direct transfer of parts from the last operation of the supplier to the using operation of the customer. It is not true that receiving inspection is eliminated for all parts that arrive at a plant. Not all suppliers attain such excellence of quality that this is a universal Japanese practice, but most material coming to an assembly plant goes to the line without inspection. The most important objective is to not have to sort defects. Therefore the quality checking is just that, sampling tests to be sure that quality of parts arriving is staying well within control. Over time, a supplier may earn enough respect that inspection is dropped; but if a defect occurs, inspection may resume for a time. It is also important not to allow a queue at inspection.

AMERICAN EXPERIMENTS TO REDUCE PIPELINE INVENTORIES

The problems with replicating a Japanese-style delivery system in the United States are obvious. Distances are greater. More material moves by rail than in Japan, and most over-the-road trucks are large and end-loaded. However, several companies have reduced pipeline stocks by a variety of approaches:

Use mixed loads, and preferably ship in truckloads rather than boxcar loads. This has required several modifications in the usual practices of traffic management.

—Use more contract trucking or company-owned trucking. Organize pickup routes of several suppliers so as to bring in mixed loads from several supplier plants.

—Load trucks in sequence so that material is removed from the truck in the order in which it is needed.

Use fewer suppliers so as to pick loads from fewer points of origin. There are a few instances in which American companies have encouraged suppliers to manufacture parts close to the point of use as is done in Japan.

Make more use of consolidation warehouses. For example, Kawasaki USA for a time had suppliers of snowmobile parts in the Minneapolis area ship small quantities to a consolidation warehouse to develop a truckload to ship to Lincoln, Nebraska. (Kawasaki no longer makes snowmobiles.)

Put more burden on the supplier to deliver by an arrival schedule rather than have the customer take responsibility for inbound traffic to their plants. Some American suppliers seem eager to do this. For example, PTC Components plans to send one truck a week to deliver timing chain to several General Motors engine plants rather than accumulate a truckload to ship direct to each plant. (However, even this modest plan encountered a snag in the beginning due to unloading delays at the GM engine plants.)

As an example of some of the American thinking, Hewlett-Packard at Vancouver, Washington, is developing a plan to have a variety of parts from a single supplier delivered in one truck trailer to their plant every Friday. The parts will be loaded in sequence on the trailer so that one day's parts at a time can be unloaded from the trailer left on their lot and taken straight to their assembly line. So far the plan is working except for refining the types of containers which need to circulate in order to make a suitable presentation of parts directly to operators in final assembly.

This example illustrates the kind of problem that will be encountered. With a week's supply in the pipeline, which now seems necessary because of the travel distance, a pull system with circulating cards or containers seems slow and awkward. The supplier needs an electronically communicated pull signal in order to stage what will be needed by Hewlett-Packard the following week. The idea is the same. The supplier will put into the pipeline what Hewlett-Packard takes out, and the companies are far enough along in process control that inspections required are minimal. (Hewlett-Packard is asking suppliers to send a copy of a control chart to show process control. Eventually they will get beyond this state.)

Another problem encountered is shifting from the traditional method of paying suppliers. Hewlett-Packard, like many American manufacturers, has been accustomed to paying suppliers according to the count by the shipment. If they are to receive more frequently, they must work out a system of accumulating a count of parts as they are withdrawn from the truck (pipeline) and pay at regular intervals. This development is also in process.

The method of withdrawing a day's inventory at a time from a parked trailer obviously has its limitations, and other methods are necessary for many parts. Kawasaki USA regards its receiving area as really a storage and staging point for material entering the plant. It is most likely that American plants will need to keep a week or so of many parts in this state due to distances from suppliers and the economics of shipping, even if many changes are made to promote mixed loads and multipoint pickups. This is the same idea as the Japanese valve warehouse. Material may even be withdrawn from this stock using a card system, as is sometimes done in Japan when daily deliveries cannot be worked out. The most important point is to keep the size of this inventory under control by regarding it as pipeline stock or intercompany work-in-process, not as stock now owned by the user. Much of this effect can be attained by paying the supplier on the basis of what is used, not on the basis of what is delivered. That stifles the practice of shipping just to accumulate another payable invoice on the part of the supplier.

Under such a system, nobody benefits from stuffing the pipeline. The supplier is forced to consider why the inventory of parts they accumulate is really needed. The system also forces the customer to give attention to the stability of the demands they place on the supplier. With a level schedule, the system will work. If it becomes erratic, the system will degenerate to the current practice.

This kind of thinking is so new in the United States that little evidence has been developed for a suitable way to go about it. A great deal of original thinking is required, so one can now only speculate how it will actually develop. Most important is to keep in mind why it is being done: reduction of overall handling costs, quick feedback to suppliers on problems, and more efficient internal operations by both customer and supplier.

IMPROVING QUALITY FROM SUPPLIERS

Consistent quality is the attribute most necessary in suppliers of companies adopting stockless production. All defects have to be sorted from the rest of the material somewhere. If it is possible for an operator to do this by a method so simple that it can be done when using the parts, then it can probably be sorted by a method so simple that the supplier can do it before the parts are shipped, so there is no reason except human error for receiving defects.

If the defects are not so easy to sort, that is even more reason to cull them before the first work center gets them. That costs money whether the customer does it or the supplier does it. The best solution is for the supplier to stop making them, but if they have been accustomed to, say, a ½ percent acceptable quality level, they may resist changing from that. In addition, if the supplier's process goes out of control for a time and that is undetected by inspection, defects can occur in streaks. That and the shipments which are too marginal for use are the disaster cases. They lead to the desire for safety stock, not just inefficiency stock. (Safety stock is some parts known to be good and kept in reserve. Inefficiency stock is the pipeline stock necessary because operations are not smooth enough to be completely synchronized.)

The objectives with suppliers, in order, are

1. Eliminate safety stocks. That means the defect rates must at a maximum be low enough that sorting operations will be effective. Entire shipments are not discarded.
2. Eliminate incoming inspection and sorting. This is possible when the production process of the supplier is good enough that there is no reason to expect defectives.

A company can begin to make stockless production effective inside its own plants provided it does not have to carry safety stocks on a large number of items, and provided it has the ability to designate assignable causes to its own process quality problems and correct them. However, the program cannot advance very far as long as a large number of sorting operations must be performed in order to allow a streamlined flow of material through the plant without a self-destructive number of interruptions.

Once the company targets a group of parts for sorting, the expense of the consistent sorting is visible enough to provoke the issue of zero defect shipments with suppliers. As companies evolve through this stage, first the sorting can be dropped, and finally the consistent sampling inspection can be dropped. Few companies will stop all forms of inspection, but will perform a random check of parts from time to time.

When an inventory for sorting is no longer needed, the company and its supplier can establish a pipeline inventory transfer arrangement if the logistics can be arranged. With inventories reduced to pipeline stock, a pull system of control can be instituted. However, a pull system will only be effective if the customer company is disciplined to work by a level schedule, which means that it must have solved its *internal* problems well enough that safety stocks are no longer needed to cover them. In the meantime, the objective is to continue to work with suppliers to eliminate inspections and work toward the time when a pull system is possible.

Working with suppliers in this way may not be much different from the way many American manufacturers work with their suppliers now. The difference is in the intensity and duration of the effort, which puts a premium on a long-term, close relationship. Establishing this relationship is somewhat like getting married. Everything is not continuously joyful, and channels of communication must be kept open to solve problems and prevent the development of unproductive emotion. It concentrates on linking the supplier's process to the customer's process, whereas inventory buffers these two processes.

The initial decision is which suppliers to work with. Some American companies hold with the same suppliers year after year, but many go with whatever suppliers appear to deliver minimum acceptable quality at the lowest price. The fallacy of that has already been discussed, but the pattern of thinking associated with it is not easily dispelled. Most American manufacturers need to reduce the total number of suppliers with whom they deal in order to form tighter links between supplier processes and customer processes. The decision on which suppliers to work with long-term hinges on whether the suppliers are willing to work toward this end.

With the increased emphasis on quality in the past two years, many American companies have already begun this. They are asking suppliers to provide evidence (such as control charts) that the production capabilities which serve them are well developed and kept in control. An extension of this is to provide suppliers with some immediate feedback useful to them. For example, samples of good parts and defective parts may be sent to the supplier for display in the area of the supplier's plant floor where the parts are made.

One of the major areas for increased cooperation is in providing correct and updated specifications to suppliers. Many American manufacturers discover that they themselves are the root cause of defectives coming from suppliers. If the customer's designers know little of the suppliers' production process, instructions on how to make the parts are easily misunderstood. The suppliers will not have correct or complete instructions on how to make the parts if the interface between the supplier and the customer for determining specifications is stiff, overly formal, and cursory. Just as design engineers are in the habit of making specifications too tight for parts made in their own plants, they may make specifications too tight on parts made by suppliers. The consequence is confusion about what is really needed, a vacillating policy on inspections, and a degree of attention to checking parts which precludes mutually improving total production process capabilities. No energy is left for it.

Just because a supplier has a good attitude about quality does not mean that the people on the supplier's plant floor know what to do to improve quality. Therefore a criterion for supplier selection is whether they have a program of process capability improvement in place. Having people in supplier plants who know statistical quality control and can use it for process improvement is a good sign. Then a system of quality feedback can be established which will result in action, not a vague promise that "it won't happen again." One of the major reasons for tightening inventory between customer and supplier is not only so that the customer improves, but so that timely feedback can be given to a supplier who has the capability of improvement.

Timely feedback from customer process to supplier process is not in itself easy to do. It requires that quality engineers, manufacturing engineers, and others have direct contact with their counterparts in the supplier companies. Those involved should know what they are trying to accomplish. Purchasing managers and the sales personnel of the suppliers should have an understanding of the kind of relationship they are trying to orchestrate, and everyone who contacts suppliers needs to understand the nature of the quality program. Even the personnel in traffic and receiving need to know, because they also communicate with suppliers. Companies must prepare internally for this, not just issue a few memos directing it.

For Kawasaki USA (and other companies experimenting with this) to make even minimal progress, it was necessary to brief everyone on how to deal with suppliers. In an American company, more emphasis must be placed on formal training and retraining than in a Japanese company. Personnel change in the contact positions at both the supplier companies and the customer companies, so it takes a program of continuous reemphasis to keep a process-to-process mode of communication going. In fact, the suspicion appears that it is very easy to underestimate the difficulty of sustaining this new form of working relationship in the American setting.

It cannot be expected that American companies will have the same kind of eternal ties that link Japanese companies. Managements change. Policies that were in effect last year in a company can be changed with a change in management this year. This entire area is one that Americans must develop in their own way.

However, quality production is the most fundamental thing to emphasize with suppliers. It is the basic from which all other progress is derived. Suppliers may wish to protest that a decreased level of defects will cost more money if they do not understand how to efficiently improve process capability, so the initial sharing of know-how with suppliers is in quality improvement if that is needed. When a review is made of which suppliers to select or retain, the best ones are those who already have excellent quality or a program in place to upgrade their process capability for quality. Next best are those who are willing and capable of doing so. Reducing the level of inventory between the customer and a supplier who understands what to do may help in overcoming some of the quality problems, but establishment of pull systems with fixed volume pipelines comes after the fundamentals are on the road to solution.

STABLE SCHEDULES AND DELIVERY SIGNALS GIVEN TO SUPPLIERS

Suppliers world-wide chronically complain about the unpredictable changes they receive in schedules and forecasts from the industrial customers. A common practice is for customer companies to give their suppliers a schedule once a week or once a month, and the schedule projects expectations for 2 to 12 months.

Readers with scheduling experience will recognize in Exhibit 9–3 the kind of scheduling and forecasting lurches that affect a substantial

EXHIBIT 9–3 Unstable forecast given by customer company to supplier company

Month in which forecast is made	*Month for which the forecast is made*											
	J	*F*	*M*	*A*	*M*	*J*	*J*	*A*	*S*	*O*	*N*	*D*
D	45	16	0	48	0	50	21	21	21	21	21	21
J	31	24	0	48	0	50	24	24	24	24	24	24
F		34	0	48	0	50	28	24	24	24	24	24
M			19	32	0	52	26	0	26	26	26	26
A				58	20	32	26	0	39	21	21	21
M					0	54	22	0	39	62	30	30
J						37	38	0	39	62	23	30
J							17	20	39	62	23	0

part of industry. *Why* an unstable situation prevails cannot be determined from the numbers, however. For example, was the situation in March caused by the unpredictable demand by the customer company's own customers, or was it caused by the customer allowing production they had scheduled in January and February to slip into March? Likewise, one can speculate about the real cause of the number changes in June, July, and August.

To understand why schedules and forecasts given to suppliers are unstable, we have to get beyond staring at the numbers. A forecast which a manufacturer gives its suppliers is usually based on what the manufacturer *plans* to produce, the nature and detail of the plan depending upon where it is in its own cycle of planning at the time the forecast for suppliers is derived from it. What the manufacturer really needs from its suppliers depends on how it actually *performs* production, so schedule stability depends on several things:

The stability of the pattern of demand coming from the market which the customer company serves.

The ability of the customer company to develop a stable final assembly schedule from this pattern of demand (and level in the case of stockless production.)

The ability of the customer company to execute a stable (and level) final assembly schedule.

The ability of the customer company to project stable final assembly schedules into a stable schedule at the points of fabrication and subassembly where material from suppliers is wanted—True whether the company pulls material directly from these operations to final assembly or whether it pushes it by schedule into an inventory from which final assembly draws. This, in turn, depends on

—Defect levels within the various production operations of the customer plant for all the many reasons which can cause them.

—Work center uptime at the customer company—preventive maintenance, reduced absenteeism, and so on.

—Setup times and flexibility.

—Other issues discussed earlier.

The lead times required by suppliers to plan action at various points in their own production processes. The longer the lead time, the less accurate will be the information upon which each action can be based.

The lead times required by each supplier and their ability to execute production well. These factors depend on the suppliers' problems with quality, maintenance, and all the rest.

Nondelivery or unacceptable quality. This snag, even if from just one supplier, can cause a change in what is being done that can reverberate throughout the entire network of suppliers.

From the above, it becomes obvious that a study of schedule stability requires a study of production processes. Improving schedule stability is more than improving planning ability; it requires improving production process ability throughout the production network, beginning with the customer company.

The campaign to improve the stability of schedules given to suppliers begins by reviewing whether the network of suppliers and the relationship with each one is capable of being stabilized. *The mix of parts or material supplied by each supplier is important.* Referring to the problems of balancing the work at work centers within a plant and to the problems of options, we can see this clearly. If the mix of parts provided by a single production facility or work center of a supplier is not one that can be held to a stable overall volume, then the schedule seen by that supplier cannot be stabilized to the degree desired, no matter how much production capability is improved by all parties.

This is one of the major reasons that stockless production companies wish to have a narrow network of suppliers. If a single part is supplied by, say, three different suppliers, then variations in actual demand from forecast may not only be amplified in the overall demand by that part but also further amplified by changes in the percentage of the requirements for the part given to each of the three suppliers.

Provided a supplier produces a mix of parts possible to incorporate into a stable schedule, they need good information at the key decision points for their own production. There are several key decision points for most:

- Future capacity decisions.
- Ordering tooling and raw material with a long lead time.
- Developing a production plan.
- Developing a final production schedule, or master schedule.
- Establishing a shipping time.

The need is well-understood, but difficult to fulfill. It is difficult to present a freshly updated forecast or schedule to each supplier at the times each one would like them for key internal decisions. They have different lead times and different internal processes. A program to provide all this to the individual needs of a large number of suppliers becomes complex.

Most customer companies can probably make improvement in the content, format, and frequency of forecasts and schedules given to suppliers without much change in the way production operates in their own plant. However, to make substantial progress, we need to consider the supplier network as a total system in the physical sense. Regarding suppliers as outside work centers, one can make a rough analysis of their processes and capabilities by concentrating on the following very general characteristics:

- Capacity—overall rate of production, capacity commitments to the specific customer doing the analysis.
- Ability to respond to overall production rate changes, short-term and longer-term.
- Ability to respond to mix changes, a function of their setup times, throughput times, defect rates—overall quality of their production competence and the nature of those processes themselves.
- Therefore the planning lead times suppliers require to make commitments themselves.

If unable to analyze this in much detail, most companies give all suppliers a uniform projection of future requirements, usually something like one or two weeks of delivery requirements (shipping releases), about one month's commitment on actual production, and two or three months' commitment to acquire material. Those with longer lead times may or may not get longer lead-time commitments.

If one had a powerful enough computer, the data base competency, and computer-to-computer linkage with a number of suppliers, a data manipulation system could probably present suppliers with the latest in automatically updated plans and forecasts at all times. The constant chatter probably would not help much if the quality of the information received is not improved, and information overload would be ignored, the nemesis of attempts to make improvement by excellence in planning alone.

To attack this problem successfully, each customer company has to begin to structure its network of suppliers so that it can be analyzed as a network of physical operations. Then they must communicate what are really new planning commitments for action as these materialize. Computer processing cycles can be short; the planning and commitment cycles of people are longer. For suppliers to revise action, the customer companies must communicate what is necessary.

Simplifying this to where it can be done usually requires simplifying the network of suppliers. Whether one develops a large number of sole-source suppliers depends on the probabilities of strikes and physical disruption of the external work centers as well as the need to simplify.

Then one must begin to work on the physical problems as well as the planning and communication problems. The more the discrepancies between plans and what is actually required can be reduced, the more that suppliers can concentrate on improving their internal operations, and the more incentive they have for doing so.

How to start this? By reducing the inventory levels jointly held between suppliers and customers, no matter who holds the inventory. The logic of development is exactly the same as the logic of developing stockless production inside a plant. As we saw in the analogy of the interval between trucks traveling in a convoy, the idea is to continue reducing the stock levels and improving both physical operations and the communication. *It is important that the customer be advanced in improving internal operations when this starts, and that suppliers understand what the purpose of the program is.*

All this works together. Cutting the inventory between suppliers and customers in turn cuts the lead time between actual production at the two locations, and that allows for action to be based on more stable forecasts of sales with less variability in schedules. Doing this means improving traffic flows, usually going to regular delivery schedules. It also requires attacking quality problems, equipment problems, and all the rest. In other words, by any logical method by which one tries to arrive at it, attaining level and stable production requires much more than just good data processing and transmission.

Supplier schedules in a stockless production industry

Using the Japanese auto industry as an example, let us focus on the scheduling of stockless production suppliers. All the OEMs of that industry follow a very similar schedule pattern. Whether this developed by plan or by historical accident is not known, but the effect is beneficial. Not all the OEMs plan and execute with the same level of effectiveness as can be established by talking with any Japanese auto supplier company that deals with the majority of them. The better of these suppliers who are using stockless production themselves know which of the OEM customers must be buffered with more finished goods inventory and which can be tracked very closely. However, all the OEMs follow a similar schedule cycle for planning and executing production. This helps the suppliers a great deal because it stabilizes their own planning to have a level schedule from which they must only make deviations in order to respond to the needs of the OEMs. This extends to building and loading parts in a sequence synchronized to the final assembly sequence of the customer.

Not all of the suppliers have programs of stockless production. Some supply items which are really commodities, items not unique to the customer supplied. Steel, fasteners, resins, and some of the paint would fall in this category. The production process for other items, such as carpet, require a long enough process time that a commitment must be made to production in advance of receiving a pull signal from the customer's final assembly line. However, most of the suppliers of items unique to each customer are somewhere in development of stockless production, or at least most of the first tier of suppliers are.

The planning and the pull signals given to the suppliers are about the same as those presented inside the OEM companies. These are developed in stages:

Forecasts. These cover a horizon of several months to several years, depending on the detail in the forecast. They enable both the OEM and suppliers to plan new models, capacity, equipment, total manpower, and raw material acquisition. Both the OEMs and the suppliers have to base commitments for steel on forecasts which are three to six months out.

Production plans. These typically extend for a six-month horizon, and are usually the basis for part and price negotiations with suppliers which are typically held every six months. The rolling plan is usually redone every month and phases into a master schedule beginning at the third or fourth month out. Two to three months out, both the OEMs and the suppliers begin planning the specific department, equipment, and manning arrangements in order to make the pull system work during a given segment of the schedule.

Firm production plans. These are typically about one month of frozen plans, which are the basis for finalizing the arrangements for making the pull system work during each 10-day block of schedule.

Delivery schedules or signals. Often done with cards, sometimes by transmission of a line sequence, these show what to ship next. It is translated into the pull signals used to initiate the pull systems within each supplier plant on this system. Commodities may be pulled from the supplier's stock with signals.

The final phase of how this planning system works with suppliers is shown in Exhibit 9–4. The days of the month on which all the OEMs transmit updated schedules to the suppliers is not identical to the day, but they all follow a very similar format.

The inquiry into why the Japanese auto industry schedule has developed in the way it has determined very little. Because the country is so small and the traffic patterns between the companies began to develop into standard schedules, the companies may have drifted into a common scheduling cycle from the realization that to have the planning out of sequence was to be at a disadvantage with the rest of the industry.

Today some of the larger Japanese auto suppliers are on-line, computer-to-computer with their major customers. The majority of the larger ones are given computer tapes updating the monthly schedules and the 10-day schedules. Some receive an updated assembly sequence daily. There is little mail delay in receiving schedules. In a system where trucks make a circuit daily, there is no reason for delays in receiving schedule information, but the telephone still gets a great deal of use between plants, particularly for explaining to suppliers what will really be needed over the next few days and why. This gives supplier supervisors a little more time to react than if they responded to move cards alone.

The basic pattern of this scheduling system is roughly similar to that which is used by *each* of the large American automotive compa-

EXHIBIT 9–4 Japanese automotive industry: Simplified overview of supplier scheduling cycle

Time of schedule receipt	*Planning horizon*	*Planning period*	*Content of schedule or directive*	
15th–20th of month M-1	3 months	Months	Month M:	Three 10-day blocks. Summaries of expected daily schedules within each 10-day block. Schedule leveled to nearly identical daily schedules within each block.
			Month M + 1:	Summary of expected daily schedule based on model and type, but no color.*
			Month M + 2:	Summary of expected daily schedule by model only.
Day N-5	30 working days	10-day blocks	Block N to N + 9:	Firm summary of daily schedule based on model, type, color.
			Block N + 10 to N + 19:	Firm summary of daily schedule based on model, type, sometimes by color.
			Block N + 20 to N + 29:	Tentative summary of daily schedule based on model and type.
Day N-5 to Day N-2	3–6 days	1 day	Firm final assembly sequence by vehicle number: model, type, color, and option detail.**	
48 hours down to 1 hour before need of delivery.	—	—	Delivery signal: move cards, containers, or actual line sequence.**	

* Models usually define frame size and major sheet metal parts. Types define engine size and drive trains (differences between sedan and sport versions, for example).

** Minor changes still occur after final assembly sequences are firm.

nies. However, that is the crucial difference. The American auto companies do not all release schedule changes at the same time. Monthly updates are released at various times throughout the month in the United States, but notice that in Japan the suppliers can rework their own plans beginning about the 20th of each month, and they will have in hand fresh updates from all major customers.

In the United States, this situation requires carrying a little extra buffer inventory in the pipelines (or parts banks, as they are called), but no one really knows how much. This covers the dead lead time in the transmission of schedules and also the dead lead time between when new schedules are received from customers when they are reworked in a regular cycle update of the supplier's own schedules.

Those who are not in the auto industry may conclude that nothing in this should really apply to them, but the auto industry is only one of the largest of those in which repetitive manufacturing prevails, and the basic principles apply to nonrepetitive manufacturing as well. Suppose a supplier has a monthly planning cycle for replanning schedules. Further suppose that the supplier has three major customers, each of which sends him one month's schedule, or one month's orders but at three different times of the month exactly 10 days apart. The supplier could choose to update his own monthly plan every 10 days. If he updates only once a month, no matter what time of month the revision takes place, a large portion of the input information will already be 10 to 20 days old and will be revised within 10 days after the replanning is complete. If the choice of update times were poorly selected, the revision could be based on inputs which are already almost a month old.

This example can be redone using shorter planning periods than a month, but the results are the same. If customers give suppliers updates at unbalanced times, the suppliers are always replanning with data that is already old. Taken to its ultimate, the suppliers are reacting on a daily basis to what are perceived to be larger changes in plan than might be true if schedule planning were more synchronized throughout an industry.

In a check of finished goods inventory of a supplier company that practices stockless production with leveled monthly schedules, the effects of schedule update cycles and mode of operation can be dramatic. If such companies are supplying customers who are themselves using stockless production, the amount of finished goods inventory on site will be only a few hours' worth or a day or two of stock at the most, depending on the shipping schedule to the customer. However, most such companies also serve customers who produce in long runs of one model at a time, and sometimes these customers want delivery of the total number of parts required for a complete run just prior to the start of that run. For those companies, 15 or 20 days of finished goods stock may be held somewhere. The supplier company must use the finished goods to buffer their own level schedule from the customers who use material in spurts, and if the schedule for the spurts is somewhat different in timing and quantity from when the material is actually used, "inefficiency" stock is required. Some customers make demands to change large quantities and delivery times on relatively short notice—well within the period of time during which a level schedule is fixed. If these demands become great enough, a level schedule cannot be stabilized and the supplier companies can only keep a large inventory on hand and try to react.

Suppliers on stockless production like to link their own production as closely as possible with that of the customers. If the requirements and patterns of operation of the different customers are greatly different, the difficulty is increased. Sometimes a supplier's production for different customers is kept separated. This opens the possibility of their scheduling production differently according to the pattern of each customer, but breaks down when the main reason for the existence of the supplier is the shared use of the same equipment for different customers. For example, Tokai Rika Company has a number of different assembly areas labeled with the names of different customers they serve, but this cannot be continued on back into their fabrication areas.

If a company regards its suppliers as work centers which happen to be shared with others, then it also shares a responsibility with other companies for the efficient management of these work centers. Industry-wide synchronization of schedules and planning cycles helps with this, but as one delves into it, the synchronization of schedules depends on making physical changes, not just changes in planning times and report formats. It begins to include a regular schedule of transport, a regular schedule for engineering changes and model changes, a pattern to planning capacity and tooling, and so on.

As its ultimate, a supplier will be so tightly linked with the customer that it will deliver unique items or options in final assembly sequence at the time required. The obligation of the customer in such a system is to send the supplier forecasts, schedule plans, and delivery signals in ever more detailed form until the point where the production of the item shipped is meshed with need for it by the customer. The concept is not much different from what is required to match different levels of fabrication inside the customer's own plant.

The ideal is to be able to fabricate a unique component in the sequence desired at the OEMs final assembly line after the latest revision to the sequence. (Even better is to have no revisions in sequence.) Several of the supplier companies are working toward that end—to have the fabrication and assembly of the part take place so as to synchronize with final assembly with no idle delay time.

In order to do this, the supplier does not have to react to a sequence totally unknown until time to start production of the item. In the case of Tachikawa Spring taking car seats to Nissan, Tachikawa has about three days' of final assembly schedule from Nissan to work from. They only have to react to the changes in this schedule as it is finally run.

Also, it is the time required to convey a part to final assembly which is important, not just distance. A light part, such as an electronic chip, could feasibly be flown half way around the world on a daily delivery schedule; but of course, it takes energy and greater time to move heavy components, so the transport planning is a function of load, distance, and congestion.

The principle of this applies even if schedules cannot be planned in as level a way as auto schedules. It is just more difficult—another stage in the development of stockless production. The ability to stabilize a schedule depends on developing the ability to mesh actual production operations together at the right time. Doing that in any operation depends on eliminating as much variance as possible from the operations themselves—attention to performance. Development of detailed schedules which cannot be executed is not the pathway.

SUPPLIER NETWORKS CONFLICT WITH ESTABLISHED CUSTOM

Because of the group orientation of the Japanese, supplier networks have grown almost naturally there. Some American companies have established supplier networks also, but in lesser degrees, and sometimes these relationships have been less than satisfactory to all parties. The idea of sole-source suppliers is often not very palatable to either the supplier or the customer. The reason is the custom of basing the price, delivery, and other conditions on negotiations, and negotiations emphasize adversary relationships, not network relationships.

There are adversary relationships between suppliers and customers in Japan also, and these exist to a certain extent among companies engaged in stockless production, but the adversary role is much diminished. This occurs because it is so obvious that everyone can only gain by forming a network of companies to compose an industry which precludes individual companies in the network from becoming overly destructive in exploiting their position in it. In addition, the culture of Japan works against this kind of development. The Japanese have no background derived from Adam Smith economics in which everyone should mutually benefit by seeking their maximum self-interest.

In the United States, dealing with suppliers is sometimes considered almost entirely a matter that can be relegated to contracts and the art of negotiation. Experienced purchasing agents recognize that smooth working relationships require much more than that, and American trade custom includes having specialists from customers or suppliers visit each other on occasion to solve problems, especially technical ones, but often even this type of assistance is derived from obligations wrested by self-interest negotiation.

In fact, it is often stated that one of the objectives of improving the schedules given to suppliers is so that the ordinary activities of delivery and quality do not consume so much time from purchasing agents that they have too little time to engage in more negotiation, or even in more search for better suppliers. Perhaps the truth is that dealing with the delivery and quality problems of the existing suppliers is more important than much of the negotiating.

EXHIBIT 9–5 Summary comparison contrasting independent suppliers versus linked supplier network

Independent suppliers	*Linked-supplier network*
1. Arm's-length negotiations by independent parties.	1. Negotiations based on operational links and mutual problems. Select a supplier to fit into a stable production network.
2. Pressure by threat of withdrawing business.	2. Pressure by establishing obligation to perform. Long-term relationships.
3. Many suppliers. Avoid sole-source suppliers because "they have too much leverage," no alternative in case of strikes, and so on.	3. Few suppliers. Develop a tight linkage with those you have. Many sole-source suppliers.
4. Purchasing agent as primary locus of communication.	4. Purchasing agent as facilitator of many points of communication.
5. Allocation of business among suppliers, based on bidding.	5. Some bidding. Mostly negotiate to develop improvements in products and processes.
6. Communication of specification by formal means. Some informal technical communication.	6. Essential specifications formally communicated. More informal technical interchange; often have technical conferences through supplier associations.
7. Prices established with suppliers. This assumes totally independent supplier operations.	7. Often mutually discuss how to reduce common costs. Discuss how the mix of parts supplied and the process performed fit into the network.

However, the practices of arm's-length negotiating and contract specification are deeply embedded in American purchasing practice. In part, this way may be caused by companies historically being situated at such a distance that all parties have to be much more specific in the start of a delivery contract to avoid the kind of communication needed to work it out as they go along. In part, it may be because Americans are very frustrated if both suppliers and customers do not have a clear statement of what they are to do. One can only speculate here. In any case, the establishment of stockless production networks with suppliers will cause many American companies to rethink their purchasing practices down to the most elementary fundamentals. For example, the most frustrating experience of Kawasaki USA in operating a small plant in the United States is that they are small while several of their major American suppliers are very large. Kawasaki, therefore, purchases only a minuscule percentage of the output of a major supplier. These suppliers typically have paid no attention to Kawasaki's overtures to create a close link which would allow frequent deliveries in small quantities. In American parlance, "Kawasaki has no leverage."

Kawasaki has gotten the same response when discussing quality with the larger suppliers. If shipping by a 1 percent acceptable quality level is the going industry practice, the supplier has no interest in doing a special sorting operation for material going to Kawasaki. The implication that doing so would help the supplier themselves establish and correct the causes of quality problems likewise falls on deaf ears. The problem is seen strictly in terms of negotiating position and price. "Who's going to pay for it?" is the response.

No one supposes that Americans are going to emulate Japanese cultural practice in supplier relationships. However, the need to form competitive industrial supply networks will force some revision of the current methods of doing business. It begins by recognizing where the primary market competition really is—for the customer of the end product. It continues by visualizing an industry as a set of linked physical processes, not as a set of negotiated power positions. This is not hard to visualize intellectually, but the patterns of daily thought which prevent it from coming about are not easily changed, to say the least.

The American OEMs who will try to form linked supplier networks will have to be very careful. The suppliers and the OEMs' own purchasing agents who cannot overcome the vision of the industry as a set of power positions will never really understand.

10

Planning and measurement of stockless production

The concept of stockless production has been presented from several different points of view. The basic concepts from limited points of view are simple, although a divergence from much current thinking. Amalgamating it all at the workplace may be a little less simple, but most of the complexity comes from factoring all the considerations together. However, that has to be done for any approach to manufacturing.

Once stockless production is in place, an overview of the planning and measurement cycle can be briefly summarized:

1. Conduct overall corporate and plant planning: markets, new products, budgets, capacity, production and transport operations, overall labor, and level final assembly and parts schedules. Some of this has already been described.
2. Determine the volume of material to have in the pipelines for the pull system, which really means planning in detail how to execute the system for a production scheduling period.
   a. Develop material movement schedules.
      (1) Transport plans between plants and from suppliers.
      (2) Material handling plans and schedules inside plants.

   b. Develop the detailed production plans inside at assembly lines and fabrication work centers to match the planned production rates, mix changes, and transport times.
   c. Check and review plans to see if they fit together.
3. Start up the pull system by plan and tighten it by observation.
4. Conduct ongoing improvement programs.
5. Measure results and look for ways to improve.
   a. Inventory status.
   b. Costs.
   c. Performance of people, work centers, and departments.

PLANNING FOR AN UPPER LIMIT ON PIPELINE INVENTORY

If the objective is to make material flow through the production network, then it is important to begin to think of inventory as something in motion, not as something stationary. That needs to be built into the planning and control of inventory. Therefore planning inventory levels needs to be related to the rate of flow of the material. This is sometimes called planning the depth of the process.

Planning should always try to relate inventory level at any point to its rate of flow, using the smallest time intervals possible so as to be able to determine the reasons for the inventory and for delays in the flow. If the inventory is moving in perfect flow,

$$\text{Throughput time} = \text{Pipeline stock} \times \text{Cycle time of use}$$

Example using an engine feeding an assembly line:

$$300 \text{ minutes} = 300 \text{ engines} \times 1 \text{ minute between engines}$$

This one is easy to see. The engine line keeps moving toward car assembly, so there is a physical flow which seldom stops. However, the idea is also used when flow is not so continuous:

$$\text{Months on hand} = \text{Balance on hand} \times \frac{1}{\text{Expected monthly use}}$$

$$\text{Weeks on hand} = \text{Balance on hand} \times \frac{1}{\text{Expected weekly use}}$$

$$\text{Days on hand}^{1} = \text{Balance on hand} \times \frac{1}{\text{Expected daily use}}$$

This can be continued until we are defining minutes or seconds on hand, if that becomes profitable. The smaller the unit of time, the

[1] In the aggregate:

$$\text{Days on hand} = \frac{\text{Sum for all parts of [Stock level} \times \text{Cost]}}{\text{Value of daily production}}$$

more continuous the flow of material that we really expect. If we are trying to promote smooth flow, we should measure at the level of detail which will reveal to us the impediments in flow about which we can expect to do something, and it is also important that the method of measurement not obscure the depth of the pools of inventory. For example, if we can use 1,000 units of an item in one day's production and we have 1,000 on hand, then we have one day on hand, provided they will be used today (or perhaps tomorrow); but not if they will not be consumed in one day, but 19 days from now. Of such issues are the arguments over methods of measurement born if the purpose of the measurement is to make performance look good without doing much to improve it.

However, increasing the speed of material flow is much more than just describing it by a flow equation. Everyone understands that if engine assembly is being synchronized to car assembly, the planning of inventory on the assembly line is only a residual of planning the operation in every other way: material attachment, material placement, material movement to the line, defective parts and their effects, inspection or test times, information flow by the broadcast, transactions to assure that the proper parts are on the correct engine, and so on. As the rate of material flow increases, it becomes more difficult to isolate the different types of operations independently. They all meld together in the complete process and have to be considered together.

Most of this discussion will describe the detailed kinds of planning necessary for inventory in a pulsed flow situation which is most feasible for much of industry given the current state of technology. However, much more development is both possible and desirable.

Having a large number of standard containers helps both the planning and execution of material handling. Stockless production companies report that the same size container can often be used for 80 percent of the parts fabricated in a single plant. Most of the standard containers (in Japan) are tote boxes about 15 inches by 24 inches and about 9 inches deep. It is amazing how many plants do the majority of material handling in boxes that size. A large number of uniform standard containers does much to standardize the sizes of material handling equipment, and that in turn promotes the flexible use of it.

Sometimes the containers are constructed to be nestable so that the empties can be stacked without consuming much space. However, if nestable containers are stacked when full, damage can result from the weight resting directly on the parts, which means having to use covers, and those create extra handling and delay. Most of the standard containers used are not nestable, and are designed to be stacked 8 or 10 high with a full load.

How many parts to put in a standard container is a function of their weight, shape, and fragility. The weight should be kept low enough that employees can move them by hand without undue strain. A set of separators is designed for each size part. By using the separators, it is easy to fill the containers with the correct number of parts without consciously counting. It is also desired not to fill the containers with too many layers because that creates a clutter from handling the separators.

Since a differently designed separator will be used with each size and type of part, the containers used for each part number should circulate between the producing work center and the point of use. For those parts which require special containers or dunnage, this is also necessary. Most of those which cannot fit into the standard containers are either very small, or they are large enough to require special containers or racks. Those are expensive, and it is important that they circulate promptly so as not to require preparing more of them than is necessary.

All these considerations are familiar in the material handling planning of many plants. What is different is that material handling and transport are typically not replanned as part of the production plan for upcoming periods. Exactly how this is done depends upon the smoothness of material flow being planned. Some of it can be done based on the number of days on hand in work in process in various areas compared with the daily volume of use. However, that is based on rough estimates, and if there is a great deal of variance in production, it can only be done by approximation. What we want is for most pick ups and deliveries to be at set times.

By physically developing the plant for a rapid flow on a regular level schedule, the size of the variances in production can be reduced. The planning of inventory then becomes not one of planning or predicting how much inventory will be held in various large pools but how it will move through the plant in a flow. As the scope of stockless production expands, this planning will extend to planning the transport schedule between plants. However, doing this requires giving some consideration to exactly how production will be performed at all points so that the plan holds together.

In addition to developing fast-flow handling of material inside plants, standard containers and material-handling methods that mesh together must also be adopted *by suppliers and customers.* Although obvious, this is an easily overlooked point in the United States, where thought is seldom given to it. Much of the savings of the system is in reducing the total cost of handling and rehandling material between point of production and point of use, and it takes physical planning to accomplish that. (In Japan, the supplier by custom fills the customer's standard container at the point where production is completed.)

Pulsed flow: How many cards for each part?

Although many variations of the card system are used, concentration on how many cards are issued for each part is instructive of the kind of planning done for each variant of pull signals. Since most parts are moved in batches in standard containers, the number of cards represents the total number of containers that must be in the system for each part, and the number of containers to flow between each point is the input to planning transport and material handling schedules.

The usual formula given for the number of cards is really an overall estimation formula that covers a variety of situations:

$$y = \frac{D(T_w + T_p)(1 + X)}{a}$$

where

- y = Number of cards; all move cards plus all production cards.
- D = Planned use rate per day, derived from the level final assembly schedule by exploding it as summarized into a daily master production schedule or by a similar method.
- T_w = Waiting time, or move card cycle time: the amount of time for one move card to make one complete circuit between the source point of the part and the use point of the part.
- T_p = The processing time *necessary* for a production card to make a complete cycle through the work center producing the part. That includes all the time for setup, running, inspections, material preparation, and the extra time necessary for the work center to maintain a balanced running operation and keep the outbound stockpoint supplied for pickup.
- a = The number of units of the part placed in each standard container.
- X = An allowance for extra time (cards) because of inefficiency. Naturally, management by policy would like to limit the value of this.

Once the number of parts per container is defined, working with the number of containers per day as a demand figure may be more convenient. Decompose this formula to show that

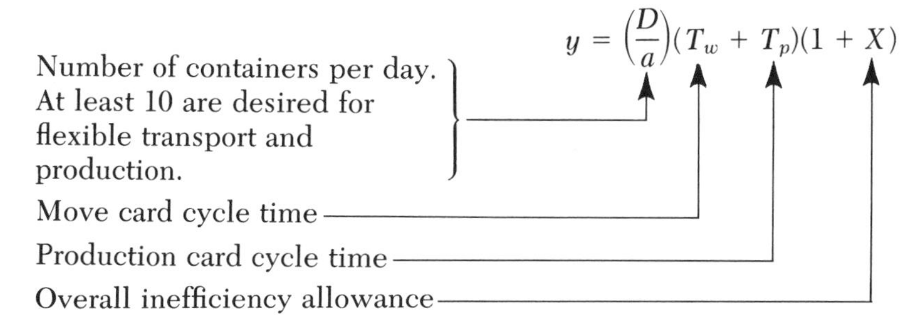

However, we may find it more convenient to analyze separately how many cards are required by the transport process (move cards) and by the production process (production cards). Decompose the formula slightly differently:

$$y = \frac{D}{a} T_w(1 + X_w) + \frac{D}{a} T_p(1 + X_p)$$

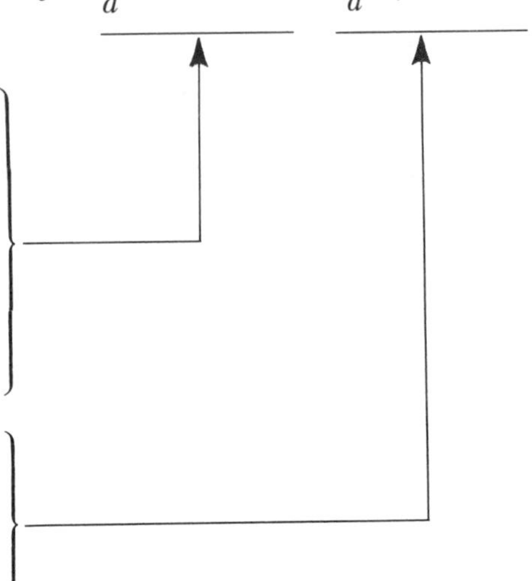

Number of move cards needed, including an allowance for inefficiency (variable times and demands). When the system is developed, we would like to keep this inefficiency allowance to 10 percent or less. That prevents the work centers from transmitting too much variance, except when mix changes may be involved.

Number of production cards needed. Because of the difficulty in determining what is really needed in processing, the inefficiency allowance for this is sometimes fuzzy to define.

Note that an overall inefficiency allowance would be:

$$(1 + X) = \frac{\text{Sum of all move cards allowed} + \text{Sum of all production cards allowed}}{\text{Sum of all cards allowed if there were no inefficiencies}}.$$

Also, if the number of cards issued is known and the rules for using them are known, various estimating rules can be used to approximate the average amount of inventory in the system from the upper limit defined by the cards, but those are only estimates.

There are several different situations for which the number of cards is planned:

1. In-plant situations.
   a. Between two producing work centers: Consider the number of cards necessary to make the part and to move it to the different points of use.
      (1) One using work center: one set of move cards.
      (2) More than one using work center: multiple sets of move cards.
   b. Parts moved from a work center to a customer.
      (1) Produce to a schedule (final assembly schedule) which moves material into stock from which it is sent to the

customer. Level the schedule. Determine the numbers of move cards necessary to bring material to this work center.

(2) Cards or pull signals come from the customer directly to the work center, just as if it were part of the customer's plant.

(3) Material goes into stock (spare parts, for example). Production control, through a stock accumulation area, issues move cards to the producing work center by a level schedule. Determine the number of move cards and a time pattern through material-handling for their circulation.

c. Draws parts from a valve inventory. The source of the parts may not be on stockless production. From the point of view of the using work center, it is only necessary to determine the number of move cards. The valve inventory itself may be replenished by methods appropriate to the source of it.

2. Parts from suppliers.
   a. Direct parts flow between the supplier and using work center; need only to allow for transit time and frequency.
   b. Flow between the supplier and the work center is delayed, usually by transaction and inspection time in Receiving. The number of cards in the circuit has to be increased to allow for the delay.
   c. The supplier makes shipments so large they must be kept in a valve inventory. If a card system or the equivalent is used at all to pull material from the supplier, it may well be a two-stage system. One set of cards from the supplier to the valve; a second set from the valve to the work center.

The planning process starts with the development of final assembly schedules and derived parts schedules, planned rates, and planned mix changes. Then those parts which are transported between plants must be evaluated to plan the transport (trucking) schedules, including transport schedules from suppliers. However, this can only be done by knowing how many containers must be transported, so unless the company is a very small one, that requires a data base and data processing to evaluate. The item masters for all part numbers have to state the container type and a quantity per each—a file kept updated by industrial engineering.

The item master (or other file) must also show the routing for each part number, the source location, and all the use locations. Unless these have been physically clarified and documented, this is not going to work very well.

Now data processing can bump these files together and derive the number of containers to be transported from point to point daily, provided they have one more piece of information—the number of move cards, or containers to be transported daily. This is the difficult part.

The number of move cards needed for the transport system of Japan can usually be calculated as twice the number of containers of each part that have to be transported with each interplant shipment, with a little inefficiency allowance. The data files have to contain a record of the allowances. Any exceptions to this general rule must also be in the file.

The number of move cards allowed for each part inside plants depends upon knowing the exact material-handling and transport procedures for each part. It is essential to plan and execute by the same set of established rules and procedures for material movement. Exhibits 10–1 through 10–3 bring out how detailed it can be, but those with experience guesstimate a lot of it. In practice, they get the system started with an approximate number of cards, then adjust by observation and experience.

A data processing file can also contain the history of the number of setups per day (or per week or per month) for each part number. Knowing the parts made at each work center, and using some general rules on allowances, a program can also make estimates of the number of production cards needed for each part at each work center. However, the updating of files being what it is, such calculations are seldom exact. It is almost impossible to maintain the data base current with the actual state of affairs in complete detail. As a practical matter, the data processing generates an estimated number of cards that may be a little loose. It may take a few extra to get a new schedule period going smoothly, anyway. It is up to the supervisors to withdraw the cards that are not needed and tighten it up as quickly as possible.

The exact method for planning this at a corporate or plant level does not seem to be established. The production planners issue a proposed plan, fairly detailed and firm for interplant transport and level product schedules, and a more tentative plan for operations internal to a plant. The supervisors inside a plant then refine the plans for their area. How much the overall plan needs to be revised, based on the detailing of the supervisors, is not a subject which many companies have wished to discuss in detail. One can surmise that it depends on how many changes in plan have been proposed and other factors that normally cause oscillations in multilevel planning processes.

Besides planning the flow operations, which are the focus here, a large manufacturer is likely to have shops which are little pockets outside the stockless production pull system. These have to be planned in other ways so that their output can be integrated into the flow, probably through a valve inventory.

EXHIBIT 10–1 Planning the pipeline stock between supplier and customer

Back in Exhibit 6–4, there were three parts being run at a total volume of 1,200 units per day. Suppose each of the three parts used one common casting and that this came from a supplier at a rate of 1,200 per day. Using the formula for the number of cards, think through what is required:

D = 1,200 castings per day.
a = 40 castings per container. The parts to be made are in run lengths of increments of 20 because 20 parts at a time go into the outbound containers. A multiple of 20 for the castings container is helpful. It allows a quick check on how many castings were scrapped, for example.

$$\frac{D}{a} = \frac{1200}{40} = 30 \text{ containers per day}$$

T_w = Assume one delivery per day. Then 30 containers per day are delivered. One set of 30 cards would be on the loaded truck coming in. About 30 more cards should be waiting for the driver to return to the foundry. That is a total of 60 cards. It takes two days for a card to make the circuit. So $T_w = 2$ days.
T_p = Production takes place at the foundry, the producing work center in this case. If the foundry is a supplier, determining how to make the parts is their business. From the point of view of the customer, $T_p = 0$.
X = The allowance for inefficiency is a matter of judgment concerning the reliability of the process. What can go wrong?
Suppose the truck can be as much as an hour late. The work center could use as many as 6 containers of castings (480 pieces) in an hour, so add 6 cards to permit extra stock at the inbound stockpoint or allow 20 percent.

The total number of move cards plus an allowance for inefficiency in transport is 60 + 6 = 66 cards.
Worked through the basic formula, the figures produce

$$Y = \frac{D(T_w + T_p)(1 + X)}{2} = \frac{1{,}200(2 + 0)(1 + .1)}{40} = 66 \text{ cards, all move cards}$$

Had the delivery frequency been increased to twice daily, the time for a move card circuit would have been halved. However, if the truck could still have been an hour late, an extra 6 cards would still be wanted. The total number of cards would now be reduced to 36, but the X factor for inefficiency would now be increased to .2—not very good.

If the system is to stay balanced, the extra 6 cards should not be present. Consider what will happen when the truck comes in an hour late and finds that the 6 extra containers of castings have in fact been used. Now the truck returns to the foundry with 36 cards, not 30, or something close to that. A big blip goes back to the foundry to replace the inefficiency stock. The action to take is to do something to stabilize truck arrivals, not add cards to the system to cover the problems. It is for this reason that Toyota and other stockless production companies do not like to have the X factors get above 10 percent.

EXHIBIT 10–2 Planning the pipeline stock between two work centers

Suppose that part P-10 is required by an assembly line which runs a level schedule requiring 500 P-10s per day. Then

$$\text{Daily demand } D = 500$$

$$\text{Container size, } a = 25$$

and

$$\frac{500}{25} = 20 \text{ containers per day.}$$

T_w = .4 day. This assumes a material handler makes the circuit five times a day very regularly. It takes .2 days to get the parts to the line and another .2 day to get the move cards back from the line, a total of .4 days. If it works perfectly, the line will have four cards (and containers) at all times, and the material handler will have four cards all the time, either attached or unattached to containers.

T_p = .25 day. This comes from the plan of the producing work center to set up P-10 four times a day. If everything works like clockwork, that is all one needs to consider for the formula. However, things may not be perfect, and in this case we may want to worry about the pickup times and the production run times not being synchronized.

Assuming that inefficiency X is zero, work it out by formula:

$$Y = \frac{D(T_w + T_p)(1 + X)}{a} = \frac{500(.4 + .25)(1 + 0)}{25} = 13 \text{ cards total,}$$

that is, 8 move cards and 5 production cards.

A little simulation check of how this will work should check our concern about whether the production runs will phase into the pickup times. Suppose that P-10 has a setup time of 4 minutes and a run time of 10 minutes per container with 53 minutes between runs. The material handler comes every 86 minutes of a 430-minute workshift.

Event	*Setup*	*Run*	*Cumulative minutes*	*Containers on hand (outbound stockpoint)*	*Notes*
Start			0	0	
Finish P-10	4	50	54	5	
Pickup			86	1	Took 4 containers
Finish P-10	4	50	161	6	53 min. between runs
Pickup			172	2	
Pickup			258	2, ok	Four containers of P-10 done now
Finish P-10	4	50	268	3	
Pickup			344	0, ok	First container of P-1 finished at 335
Finish P-10	4	50	375	4	
Pickup			430	0	

In reviewing this, note that with only five production cards, the work center could not have placed six containers on hand with the second setup. Likewise, they could not have had four containers of P-10 complete for the pickup at the 258 mark. It would take six production cards to do that.

EXHIBIT 10–2 *(concluded)*

As long as only four setups per day are planned for P-10, that will be the result. If they can manage five setups a day and synchronize with the pickup times, they could become perfect. As it is, they need six production cards just to meet a schedule that has no slack for trouble.

How many move cards? If the material handler makes a round robin trip to the assembly line every 86 minutes, it takes 172 minutes for the round trip. He has to leave four full containers each time and pick up four move cards to take back. How many cards are needed depends on exactly the procedure used.

If the assemblers remove the cards as the containers are used, then the system works exactly as the trucking system: leave four cards, pick up four cards for a total of eight cards required. Perhaps he would like an extra card just in case his timing is a little slow getting to the line—nine cards. However, a better way is to have enough slack in his schedule that he should not be late but always be able to arrive a little ahead if necessary—eight cards.

On the other hand, if the material handler takes the containers to the line and feeds them into a rack from which assemblers will withdraw them for use, he only needs four cards and no extras for bad timing. He pulls the cards himself as he feeds the containers into the rack.

That not only allows the withdrawal of cards to get a little out of phase with the actual use of the parts, but the material handler is also pulling the same cards off the containers that he just put on them at the source work center. It does not take a material handler long to conclude that he does not need to do that. In fact, why have either the assembler or the material handler pull cards off the container at the assembly line? The material handler has to return the right number of empty containers anyway. Leave the cards on them. Pull the cards and replace at the source stockpoint.

Anyway, determining the number of move cards depends on knowing the timing of the pickups and the exact procedure used. If the material handler were able to get more parts five times a day but it only took 10 minutes to make the trip and restock the line, only one container would have to be on hand at the time he left the line. That takes five cards.

The moral of this is that we want to establish a material-handling schedule and procedures that support a level schedule. Within that, we want to give no inefficiency allowance to the material handlers unless it is unavoidable. The extra cards sloshing through the system contribute to more imbalance between operations than may be necessary.

Why not leave the move cards on the containers that are left at the source stockpoint and use no production cards? In effect, the work center just fills up the empty containers. That is done, and is the prevailing practice at some companies, but it pools the inefficiency allowance of the source work center, the using work center, and the material handlers. It is just a little harder to isolate the source of problems and compress the variance in production. Keep in mind that the use of the cards is to identify variance, so how tightly it is set up depends on how much can be attacked and held. Cards are used for a purpose, not to impose a burden for the sake of observing a formality.

EXHIBIT 10–3 Running a balanced plan at work center A

Consider the same work center and three parts that were described in Exhibit 6–4. It runs three similar parts, and the plan is summarized as follows, assuming an 8-hour shift and 430 available work minutes per shift:

Part number	*Required per shift*	*Container size*	*Containers per shift*	*Setup minutes*	*Run time (minutes/ piece)*	*Run time (minutes/ container)*
P-1	500	20	25	5	.25	5
P-2	400	20	20	5	.30	6
P-3	300	20	15	10	.40	8
Totals	1,200		60			

Daily plan: three setups for each part per shift.

Possible number of containers for each setup

	First	*Second*	*Third*	*Total*
P-1	9	8	8	25
P-2	6	7	7	20
P-3	5	5	5	15

The following simulation (just as in Exhibit 6–4) shows how it might work:

				Balance on hand—outbound stockpoint (number of containers)			*Taken to final assembly (number of containers)*		
*Event**	*Setup*	*Run*	*Cumulative minutes*	*P-1*	*P-2*	*P-3*	*P-1*	*P-2*	*P-3*
Start			0	5	7	5			
(9) Finish P-1	5	45	50	14	7	5			
Pickup			86	9	3	2	5	4	3
(6) Finish P-2	5	36	91	9	9	2			
(5) Finish P-3	10	40	141	9	9	7			
Pickup			172	4	5	4	5	4	3
(8) Finish P-1	5	40	186	12	5	4			
(7) Finish P-2	5	42	233	12	12	4			
Pickup			258	7	8	1	5	4	3
(5) Finish P-3	10	40	283	7	8	6			
(8) Finish P-1	5	40	328	15	8	6			
Pickup			344	10	4	3	5	4	3
(7) Finish P-2	5	42	375	10	11	3			
(5) Finish P-3	10	40	425	10	11	8			
Pickup			430	5	7	5	5	4	3

* Numbers in parentheses show containers made exactly according to plan.

However, this is not following a pull system. We have put no upper limit on the amount of inventory in the stockpoint.

Note that we have used 15 containers for P-1, 12 for P-2, and 8 for P-3. That is, we would use that number of production cards sometime during the day.

EXHIBIT 10–3 *(continued)*

Estimated number of production cards:

	Part		
	P-1	*P-2*	*P-3*
Maximum run length for balance	9	7	5
Maximum pickup quantity	5	4	3
Total	14	11	8

This is only an estimate for this kind of situation.

	*Event**	*Setup*	*Run*	*Cumulative minutes*	*Balance on hand—outbound stockpoint (number of containers)*			*Taken to final assembly (number of containers)*		
					P-1	*P-2*	*P-3*	*P-1*	*P-2*	*P-3*
	Start			0	5	7	5			
(9)	Finish P-1	5	45	50	14	7	5			
	Pickup			86	9	3	2	5	4	3
(6)	Finish P-2	5	36	91	9	9	2			
(5)	Finish P-3	10	40	141	9	9	7			
*(5)	Finish P-1	5	25	171	14	9	7			
	Pickup			172	9	5	4	5	4	3
(6)	Finish P-2	5	36	212	9	11	4			
(4)	Finish P-3	10	32	254	9	11	8			
	Pickup			258	4	7	5	5	4	3
(10)	Finish P-1	5	50	309	14	7	5			
(4)	Finish P-2	5	24	338	14	11	5			
	Pickup			344	9	7	2	5	4	3
(6)	Finish P-3	10	48	396	9	7	8			
**(5)	Finish P-1?	5	25	426	14	7	8			
	Pickup			430	9	3	5	5	4	3

* At this point, the balanced plan called for eight containers to be made of part P-1. However, the pickup had not yet been made, so there were only five unattached production cards. The run stopped one minute before pickup.

** At this point, no more P-1 is authorized because pickup has not yet been made. If they follow the rules, they will set up next for part P-2, but the shift is almost over. P-2 is the only part they are authorized to make at this particular time.

Although the demand pattern from the assembly line was perfectly level and the material-handling pickups worked like clockwork, the work center's day deviated from the planned balanced run lengths. The results:

	Part			
	P-1	*P-2*	*P-3*	*Total*
Containers made:				
Plan	25	20	15	60
Actual	29	16	15	60
Setups made:				
Plan	3	3	3	9
Actual	4	3	3	10

EXHIBIT 10–3 *(continued)*

For those who like to play games, note that the number of containers for part P-1 never dropped below four in the outbound stockpoint. Reduce the number of cards for P-1 from 14 to 10 and work through the arithmetic, starting with a balance on hand of 1. Nothing changes, at least on this particular day.

Now reduce the number of cards for P-2 to eight without changing the beginning balance. The first run for P-2 stalls at the 61-minute mark because there are no more cards authorizing production.

If we try this for both P-1 and P-2 at the same time, the entire work center stalls out at the 75-minute mark. What we are doing is forcing more setups and shorter runs, but at the current setup times we do not have the capacity for it.

The number of production cards has to reflect not only a plan, but expected changes in plan. Suppose we expected the mix change to switch back and forth from P-1 to P-3, in other words:

P-1 containers per shift: 15–25
P-3 containers per shift: 15–25

Mix change expected:

	Part		
	P-1	*P-2*	*P-3*
Estimated number of cards:			
Maximum run for balance	9	7	9
Maximum pickup quantity	5	4	5
Total	14	11	14

	*Event**	*Setup*	*Run*	*Cumulative minutes*	*Balance on hand—outbound stockpoint (number of containers)*			*Taken to final assembly (number of containers)*		
					P-1	*P-2*	*P-3*	*P-1*	*P-2*	*P-3*
	Start				5	7	5			
(9)	Finish P-1	5	45	50	14	7	5			
	Pickup			86	9	3	2	5	4	3
(6)	Finish P-2	5	36	91	9	9	2			
(5)	Finish P-3	10	40	141	9	9	7			
	Pickup			172	6	5	2	3	4	5
(8)	Finish P-1	5	40	186	14	5	2			
	Pickup			258	11	1	OK*	3	4	5
(8)	Finish P-3	10	64	260	11	1	5			
(8)	Finish P-2	5	48	313	11	9	5			
	Pickup			344	8	5	0	3	4	5
(10)	Finish P-3	10	80	403	8	5	10			
(3)	Finish P-2	5	18	426	8	8	10			
	Pickup			430	5	4	5	3	4	5

* This pickup would be covered because the run of P-3 is nearly finished.

The work center reacted quickly to the changed mix in the pickup, and they altered their setup sequence to adjust to it. At the end of the shift they are covered on all three parts for the next pickup, which gives the next shift 172 minutes to get through a setup on all three parts, which is possible, but not good enough. They are short production for the shift. The results:

EXHIBIT 10–3 *(concluded)*

	Part			
	P-1	*P-2*	*P-3*	*Total*
Containers made	17	17	23	57
Setups	2	3	3	8

Since they are behind in total output, it will be a little harder to stay up with the line on the next shift, and they are not going to be in a good position if they are still behind and the mix changes back. Since P-3 has a longer run time than P-1, the total work day has to be a little longer to allow for this—about 30 minutes longer. The shift really needs to stay a while and get themselves covered through two pickups on one of the parts and catch up on total output at the same time. Extending the last run of P-2 is one way to do that.

How many move cards will be needed for each of the three parts in this situation? Assume that we use the round robin method of material handling, with no allowance for bad timing on the part of material handling. Then the number of cards for each part will be twice the number of containers which are picked up on each material-handling round. Since that number can change because of mix changes, the number of cards issued for each part has to be large enough to allow for that. The number of move cards are shown below:

	Part			
	P-1	*P-2*	*P-3*	*Total*
No mix change expected	10	8	6	24
Mix change expected	10	8	10	28

The ability to react to mix changes is bought at the price of permitting some extra float in the system, and that can work against the effort to uncover problems by tightening the inventory.

Inventory (work) planning by supervisors. The planning done centrally is based on historical data and estimating methods. The supervisors, with staff help, then detail their plans. This planning follows a pattern which is more universal:

1. Work from the cycle times of use for all parts as derived from the final assembly schedules (or final parts schedules in the case of parts made for other companies and spare parts.)
2. Establish the work cycles for all parts produced by a work center. These must allow for everything foreseeable.
   a. The number of setups per day which can be done for each part, considering the current setup times.
   b. The points at which inspections must be made and how they must be made. Hopefully, all inspection can be done automatically or built into the production operations; but, considering the current status of quality, this must be established so as to prevent defects from entering the area, being made, or leaving the area.

   c. The work cycle times for each part in running operation, allowing for material preparation, tool changes, quality check—interruptions for activity which cannot always be done at setup times.
   d. The time needed for preventive maintenance, making observations, and generally maintaining an orderly work area. (Although everyone wants to have equipment running in top shape and do as much PM as possible between shifts, current equipment problems or other problems must be considered for a given schedule.)
   e. Consideration for the problems of maintaining balanced operations in a pulsed flow, as discussed in Exhibit 10–3.[2]
3. From these considerations, each work center will have a work plan, only one part of which is the number of production cards for each part made there. The other major parts of it:
   a. Number of setups per day for each part; run lengths for each part.
   b. Expected regular times for picking up materials needed by the work center; expected times completed parts will be picked up.
   c. The quality plan: inspections, check points, and times—the rhythm of quality control procedures and checks.
   d. The plan for making observations about equipment or processes; preventive maintenance activities.
   e. Plans for making improvements in equipment or procedures.
   f. Plans for product changes; trial runs and such.
   g. Operator schedules and assignments: expected overtime, and rotation of assignments for training and flexibility.
4. After a basic plan is formed for all production work centers, the material movement plan inside the plant may be revised for better balance of the work centers, to provide more level loads to the material handlers, for example. The proposed plan should have already considered changes and improvements in the methods of material movement, but the supervisors may have more detailed changes to try. At any rate, the number of move cards to issue for each part will be based on this plan by the supervisors. The material-handling work plan will be similar in nature to those for the production work centers. Revisions to the material-handling plan should cause only small changes in the number of production cards issued at the production work centers.

---

[2] The data presented for the work centers in Exhibits 10–2 and 10–3 are constructed to illustrate the points made there. The work cycles as shown do not make sufficient allowance for the considerations presented above.

In Japan, the line foremen do much of this planning personally. The department supervisors may assimilate their departmental plans together personally, and the supervisors may have an assistant. The line managers can ask professional staff for assistance with the more difficult planning problems.

There are advantages in this. People are more likely to understand and follow their own plan. Line supervisors also do not have time to overplan, and trying to preplan in too much detail only results in a plan which still has to be adjusted anyway. Supervisors and staff learn which foremen need great help in planning and, therefore, training. Everyone learns by experience when a plan has been developed enough that it should be attempted. The entire process puts great emphasis on developing the people who are closest to the action, and on executing the plan—not elaborate, detailed planning remote from the action. Not the least of the advantages of supervisory planning is that it prevents the ballooning of staff to do the job.

Workpace is an important consideration in work center planning. A plan which calls for more setups than the equipment modifications are ready for can be physically demanding. Supervisors and foremen—the planners—need to keep in mind that the long-term trend in operator activity is to keep them busy enough to stay alert and thinking. Manual labor is to be removed from the tasks, not increased in pace. As the complexity of tasks increases, workers need enough slack time to correct problems as they occur and to react to changes in the pull signals, and the more automated the work center becomes, the more this is true. Again, the closer the planner is to where the work is accomplished, the more likely he or she is able to exercise good judgment about this.

Note that the formulas and the system of planning are directed throughout to making the planners examine *why* inventory exists. Good performance really depends, however, on the managers and supervisors staying close enough to the operations to know if foremen are tightening the system or allowing it to have more slack than necessary. There is no way to avoid the diligence of management in making any system work.

THE CHANGEOVER BETWEEN SCHEDULES

Major changeovers between periods of planned schedule usually occur about once a month. Minor changes may occur about every 10 working days. Several of the activities which take place at these times have already been described: preparing equipment, tools, material and layout for the new schedule, changing the operator assignments; counting the stock in the pipelines; producing samples or trial parts;

accumulating and updating the records; and introducing engineering and model changes. These are also favorite times for reworking the production areas to try out new improvement ides. The amount of dead production time depends on the amount of change activity, but for a plant making major changes on the floor these periods can be hectic. Areas such as U-lines can be completely revised.

The changeovers are also major milestones in the march to more automation: times to implement the changes in system and procedures. Times in between changeovers are for tightening the system, observing results, and preparing for new changes.

One of the troublesome regular activities of changeovers is restocking the pipelines with new parts. A final assembly line cannot pull parts from work centers which have just started making them, and this effect continues all the way back to raw material. The changeovers must start at the lower levels of production and bubble up to final assembly.

Suppose parts are delivered once a day between work centers at the different levels of production. Then the work centers three levels down from final assembly are three days behind. Intercept the old cards at the lowest levels three days before the change at final assembly. The lowest level centers make their changes and stock up for the new move cards to arrive. Next day, the next higher level work centers have material to draw as they make the change and start on the new schedule and stock up, and so on up to final assembly.

If material is moved more than once a day, the lowest level can start three move cycles behind final assembly; but that assumes that changeover will not take very long, and all of a plant is likely to be shut down at once with staggered stopping and starting beginning at the lowest levels.

One way the old parts for old models are cut off is by cumulative counts, a system used by auto companies throughout the world. A cumulative count is made of the number of major parts produced, and this is compared with a cumulative count of the parts that will be needed through the last unit in final assembly. That system usually works well in cutting off the old parts even if the flow is not smooth and has large inventories in it.

The smoothness of the start-up on a new schedule period depends on how well the preparation for it has been done. A large number of changes in production processes, combined with a large number of product changes, certainly requires more debugging, and it is the function of management to keep the changes in executable increments in refining the system.

The engineering or model changes that are not too large in scope are tried and practiced before the changeover in which the changes are made part of the regular schedule. The motorcycle companies have many model changes to manage each month. By trying these changes at the ends of shifts as one schedule smooths out, the supervisors and workers can be better prepared to execute them in the production plan they prepare for the next schedule.

Automotive annual model changes are too complex to be treated this way. A "flying" change may be possible at final assembly, but not at heavy fabrication even after great advancement in setup times and flexibility. However, learning to manage the changeovers well is a part of making total progress with the system.

One company, Yamaha Motorcycle, has a large number of model changes, and they basically use the card method for their pull system. Of their 450 or so models, about 25 change every month, and many of the parts are not common between models. They have been developing a system they call PYMAC. One of its major purposes is to maintain running totals of major, unique parts in the pipelines. The move cards are printed with bar codes used for making counts. That enables them to make model changes and complex engineering changes with less difficulty, sometimes on the fly. It is a complex system. (They have 200,000 part numbers in it, the numbers derived by adding a suffix to the regular number so as to designate source point and use point.) This conflicts with the desire for simplicity in systems, but Yamaha thinks the price is worth it because the motorcycle market demands shorter and shorter product cycle times.

Learning to execute the changeovers well is part of the total process of learning to operate with stockless production. A period of fixed schedule cannot be executed well if it takes too long to get it started. If transitions from schedule to schedule go smoothly, the operations are potentially more flexible, and perhaps the length of fixed period schedules could be shortened.

MEASURING RESULTS AND PERFORMANCE

Considering the nature of the production system, the kinds of data that should be kept at a typical stockless production work center or department can be briefly summarized:

1. Data captured on a daily basis.
   a. Number of setups completed, per machine and per part.
   b. Counts of parts and units completed.
   c. Defect and scrap counts, categorized by cause.
   d. Quality control data.
      (1) Control charts of process capability checks.
      (2) Measurements taken to plot trends.
   e. Material changes (when relevant).
   f. Equipment malfunctions, categorized by cause, if known.

   g. Tool and attachment changes and usage.
   h. Hours worked (clock hours).
2. Data captured less frequently than daily, but on a regular basis.
   a. Setup times and procedures, by part and by machine.
   b. Work cycle times and content.
   c. Inspection points and procedures.
   d. Specifications of parts.
   e. Work instructions.
   f. Data on problems.
      (1) Suggestions and proposals.
      (2) Improvements made.
   g. Worker skills and qualifications demonstrated.

All of this is needed by the people operating the department in order to function by stockless production. A great deal of data is collected in some way, but much of it must be done in a hurry and therefore done very simply. Not much of it may be transmitted to a central data base, except in summary form.

The principle of collecting data is to collect that which is necessary for taking action. If the people who need to use data to improve operations also collect it, they understand the reason for its collection and something of the validity of the data. Most of the corrective action and improvement takes place at the department level, so data collection is centered there. This is important when increasing the automation and flow rate of production. The analytical work (staff) has to stay close to the action with as much of it as possible embodied in the workers and line managers, a policy which follows from the premise that the workers are to become more and more the controllers of processes and equipment, not just performers of motions.

Measurement is essential, but it is also an operation just like any other production operation. It takes time. Even if it is done by indirect labor, it still takes time. Regarding the measurement and collection of data as part of the production task itself keeps this in perspective.

It is also useful to keep in mind that measurements are taken to improve and control a flow of production, not slow it. Holding inventory in batches to count it or measure costs is contrary to this. Cost and inventory systems for stockless production are designed for a flow, and the data is captured from a large volume of quick and simple transactions. Performance measurement is intended to promote improvements in flow.

Inventory systems

This description is based on the typical Japanese stockless production company. Inventory systems are basically what Americans sometimes call four-wall systems. That is, a recorded transaction does not usually distinguish when material moves from raw material into work in process. Raw material is distinguished from work in process at the time of the monthly count, depending on whether it is in the work center areas or held as valve stock in receiving or staging areas. Most finished material is counted as finished goods when the final count is made leaving assembly.

A receiving count is made of most items coming into the plant even if these bypass the receiving area and go straight to the point of use. This is common in assembly and subassembly operations. Usually there are terminals in the line staging areas to record incoming material.

Once stockless production has advanced to where most suppliers to final assembly can ship zero defects, no inspection is required. However, most plants have a few items that are exceptions, even in final assembly, and the majority of material for fabrication may be inspected. However, the defect rates are low and rejections are rare, so the inspection area does not backlog with the tough decision cases. Much of the inspection appears to be for the purpose of feedback to the supplier as well as for screening what enters the plant.

One notion of safety stock for quality rejects differs from American practice. One or more standard containers of parts that are known to be good are set aside. Parts can be withdrawn to cover for those that turn out to be unusable, provided that only occurs occasionally and not in streaks. Checking how many have been withdrawn from the safety stock is a double check on how many defects have slipped all the way into the production area. One can imagine that such a safety stock is a point of discussion with the supplier.

There usually is not space in the work areas for bulky material or for items delivered every other day. Steel coils, bar stock, and similar items are usually seen in staging areas or valve inventories. Smaller items can often be held in the receiving area, or near it, and drawn as needed by the card system.

A stockroom may be used for maintenance and repair parts, and this is generally a part of a maintenance or tool-and-die area. The author does not recall seeing any production material in a stockroom of a company advanced in stockless production. Many small hardware items (commodities) are really ordered by an order point system, but not in large quantities by American standards. Upon arrival, they are usually put into a staging area where they are kitted to take to the work areas. Most of these items are used on assembly lines. For material handling purposes, they are doled out in small quantities, and with level schedules the ordering of them is a very regular, repetitive process.

The labeling and control of small commodity items is more a concern for keeping the assembly areas uncluttered and the assembly process efficient than it is a concern for keeping an accurate count of the overall quantity on hand. In effect are several visual two-bin methods, such as "get another container when you can see the ring on the inside of the one from which you are withdrawing." The actual methods of attachment of small parts seem very comparable to American assembly methods, but there is greater attention in both equipment and procedure to metering out small parts so that they are easily selected as needed and work areas are kept organized.

With so much emphasis on physical organization and a pull system for getting the right part to the right place, inventory accuracy is not a subject of great discussion. Inventory accuracy is important when it is necessary to have a large inventory out of sight, and planning decisions must be based on the inventory levels themselves. In the United States, inventory accuracy generally connotes having the record correspond exactly with the physical status. To do that, the inventory system has to maintain accurate input-output transactions, and that is the purpose of the techniques for improving inventory accuracy.

With almost all the inventory out in the open and the flow moving quickly, there is no less emphasis on accurate transactions. However, it is easier to avoid mistakes because there are visual reminders which cause trouble very quickly if a mistake (in identification, count, and so forth) is made.

Many machines have counters, and, in addition, the standard containers and regular rhythm make it easy to keep accurate counts on the production floor. Counts in receiving are aided by scanning bar codes on the cards and by using forms that can be optically read. The most difficult accuracy problem reported is in maintaining the balances of items sent out to subcontractors and returned. If the subcontractors work by stockless production themselves, the task is much easier, but the problems occur from exactly what you would expect—subcontracted assembly operations with mix changes and also fabrication operations with scrap.

Scrap reporting is not a problem, even in operations where a high scrap rate is still occurring. With a quality system in place to study the causes of defects, the scrap made is generally reported and analyzed in the department which makes it without the stigma associated with it if scrap is a negative factor in bonus pay or advancement.

At the end of the month, the reconciling of counts and inventories is very straightforward:

(Cumulative receipts for the month M) −
(Usage by backflush, other usage, and cumulative scrap) =
(Inventory for month M) − (Inventory for month M-1)

The inventory count is taken at a time when everything is stopped and before any department starts changeover. The backflush is a deduction of a bill of material explosion based on counts of units made, or on counts at major count points in the plant. The entire procedure is very similar to that used by many repetitive manufacturers in the United States. None of the companies seemed to have enough trouble with reconciliation of the inventories to merit any specific discussion.

Cost accounting

All the Japanese companies involved in this use a form of process costing throughout. A process-costing system kept on a monthly cycle certainly should not be difficult, given the data kept by the departments and work centers. There are no work orders except for special maintenance jobs, charged to departments or work centers.

Most accounting is based on the cumulation of many transactions over a month. Suppliers are paid on the basis of the cumulative parts received over the month, generally by about the 15th of the following month. If the schedules are level, the amount paid to each supplier can generally be anticipated in advance.

Although questions were asked, Japanese do not seem to set up any special accounts to capture costs unless they are involved in special studies of particular problems. It is apparent that the cost accountants, as well as other staff members, really serve the needs of the management and especially the department managers.

They observe what the author dubbed the Heisenberg uncertainty principle in costing. That is the principle from physics which says that attempts to measure what is happening may distort what is happening. Therefore costing is mostly a matter of developing unit costs each month based on the end-of-month data, and the management watches the trend in these.

No company wanted to say how they developed any form of standard cost upon which pricing decisions are made, or on which expense budgets are based. However, several companies referred to target costs. These appear to be unit costs which should evolve from an expected trend line resulting from improvements. Department managers are conscious of those. They receive expense budgets and target costs. First-line supervisors typically do not have direct budget responsibility. Results are measured by the costs as based on end-of-month cumulative totals, incorporating all the costs of the department.

The work cycle times do not appear to be used in costing. Supervisors and department managers keep these within the department, and they are used for measuring their own progress—an industrial engineering methods-study use of them, not a cost-accounting use. In stockless production, the notion of standard allowances from a basic elemental time does not appear. Results are based on total departmental output and resources used for the month.

The concept of this is important, more important than exactly how the numbers go together. Actual costs as approximated at the end of the month reflect relatively constant overhead allocations. People may concentrate attention on a particular operation within a department one month, very little the next. The system does not create illusions with little shifts in how overhead is allocated to direct labor, for instance.

A company which has long been accustomed to costing by allocation to direct labor may need to rethink its system. How would costing be done if the operations were totally automated and there were no direct labor? If stockless production is successful over the long term, that is what should happen. The direct labor base will continue to shrink so that a different means of allocating cost must be found.

Performance measurement

Under stockless production, the major performance measures emphasize the number of processes and practices changed to improve material flow and reduce labor content, and the degree to which they do so. If the processes physically improve over time, lower costs will follow.

In a Japanese stockless production company, if any distinction, however small, is made in pay based on performance, it is based on departmental performance, and a department contains at least 50 people. Any such distinction is based on trends which have been established over a year or so and which seem likely to be permanent changes. This puts improvement of activity as a top-priority goal. It also encourages improvement as a complete operation. Japanese culture may have a bearing on why this was done, but the method of performance measurement seems likely to create a broad impetus for change in any group.

A department head is likely to be evaluated on the following:

1. Improvement trends.
    a. Total improvement projects and total implemented.
    b. Total improvement projects and total implemented per capita.
    c. Trends in costs or toward target-cost goals.
    d. Productivity:

$$\frac{\text{Measurement of department output}}{\text{Total employees (direct + indirect)}}$$

2. Quality trends.
   a. Reduction in defect rates.
   b. Improvement in process capability.
   c. Improvement in quality procedures.
3. Running to production schedule or plan.
   a. Attaining level, linear line schedules.
   b. Providing parts whenever others need them.
4. Trends in departmental inventory levels—speed of flow, for example.
5. Staying within expense budget.
6. Development of the work force.
   a. Increased skill evidenced in their work.
   b. Versatility in different jobs.
   c. Participation in changes.
      (1) Participation in training.
      (2) Number of suggestions made and percentage implemented.
   d. Morale:
      (1) Absenteeism, complaints, and so forth—negative indicators.
      (2) Enthusiasm—positive indicator, subjectively evaluated.

Except for direct budget responsiblity, the same factors are likely to be used in evaluating the performance of first-line supervisors. This is certainly a more complex set of measures than just evaluating performance based on a few overall numbers. Executives have to be familiar with what is happening in production in order to use them wisely. They also discourage supervisors from deferring attention to some parts of their job in order to appear proficient by only one measurement. They encourage preventive maintenance, for example.

One measure common on most industry is not in the list—utilization. That depends on the schedule the department is to run and on what they are doing to improve equipment. As we observed earlier, it can be a deceptive measure, and, for a time, one of the objectives (U-lines) can be to make the machine operations dependent on operator presence.

These measures take a long-term view. A performance measurement which is biased toward only one short-term indicator is not likely to do that. For example, if performance measurement is heavily weighted on efficiency, a supervisor may not be vitally interested in anything more than this month's efficiency rating. That creates incentive to defer maintenance, avoid setups, and not stop a process which is marginal on quality. (Two of the first experiments with these ideas in American plants met an early end because the supervisors did not dare risk their efficiency ratings in order to do development work.) Most companies rate their supervisors on much more than efficiency, but it is where the supervisors themselves perceive the emphasis that counts.

A similar result can be expected from any performance measurement system that is oriented only to keeping the current form of production running. A set of performance measures oriented to long-term revision of the production process greatly upgrades what is expected of production supervisors. That in turn greatly increases what the supervisors expect in the long run from those who work for them. It creates a great deal of attention on the production processes themselves, which is what is wanted.

11

Implementing stockless production

Only the initial objective of stockless production is the development of a pull system and all that is required to make one work. Beyond that, the objective is to develop into full automation and thereafter continue to improve upon that. However, each company must start with the development of a basic pull system, and it is far from easy to determine how to implement that. The degree of difficulty depends upon the status of the company's production when it starts. It will be less a change in thinking if a company already has repetitive manufacturing in a substantial part of its operations than if its plants, currently operating by job shop methods, can potentially be operated by repetitive manufacturing.

It is expected that some managements will only wish to borrow a few ideas from stockless production but not undertake a rigorous and far-reaching program. Others will wish to promulgate a revolution. Assuming that management is contemplating the revolutionary scale program, there appear to be two major milestones:

1. Making the full commitment to go ahead with it.
2. Getting the plant physically converted to a pull system of planning and control, from which further development can proceed.

The companies which have gone this route before testify to the great fear of starting, because, once a full commitment is made, no aspect of the production operation is unaffected by stockless production. Carrying out an integrated conversion to stockless production is difficult. If an operating manufacturer tries to change everything at once, so many subprograms are in progress that no one can keep it coordinated. That is very true, and the program in its infancy is very demanding. (See Exhibit 11–1.) Management must be determined to lead people through an intense period of confusion while they adapt to new ways of thinking. There is a hazard that management can become content with one or two dramatic accomplishments which are only a small part of stockless production and stop far short of attaining a level schedule with a pull system in place.

Assuming that management has made a full commitment to a complete program of stockless production, the following discussion is based on the prospect of a single-plant manufacturer, or a manufacturer with only a few plants. It does not address the broad strategy issues of a large corporation. That is left for the last chapter. The more specific programs of implementation are based on a synthesis of Japanese implementation history, with a brief history of trial projects in the United States. Most of the experience is Japanese, and therefore one can only extrapolate the problems to be expected from what little has so far been observed on American soil.

The actions to be taken can be classified in two ways:

1. Physical changes.
2. Organizational changes.

This may occur in roughly four phases:

1. *Conceptualization:* Learning, devising strategy, planning, experimenting, and developing confidence.
2. *Preparation:* Revising the plant, reducing setup times, improving process capability, and improving plant housekeeping to a point where the conversion to a pull system is physically possible.
3. *Conversion:* Changing from whatever method of material control now prevails to a pull system for the entire plant.
4. *Consolidation and continued improvement:* After the plant is basically operating by a pull system, much further improvement is possible, and the system can be the basis for evolving into full automation by whatever technological means seem feasible.

Naturally, any real program may not be done in such a systematic way. Every plant has improvement programs operating all the time. It may well be, for example, that a program to reduce setup times could start with little forethought while additional action is still being planned. There is no such thing as a completely detailed plan for a revolution.

EXHIBIT 11–1 Summary of the nature of different jobs with repetitive stockless production

Workers:

1. Broad range of tasks:
    a. Do some setup. The job may evolve into almost all setup.
    b. Monitor quality directly. Stop the process for a defect.
    c. Provide information or signals to others:
        (1) For material needed.
        (2) For machine or process problems.
        (3) For quality problems.
    d. Organization and housekeeping of area.
    e. Routine preventive maintenance.
    f. Material handling in area.
2. Involvement in improvement.
    a. By direct suggestions or involvement in improvement groups.
    b. By contributing to the work of others who are developing improvements.
3. Maintain flexibility to do several jobs well. This accords with having a limited number of worker classifications, and with a great deal of cross-training.

Foremen:

These employees perform daily assignments and record keeping on those working for them. (If they also do all the other work shown below, they cannot supervise a large number of people.)

1. Provide on-the-spot coordination and leadership for the improvements being made in the department:
    a. Housekeeping.
    b. Organization and layout.
    c. Equipment modification. (Projects for setup reduction and process capability improvement.)
    d. Material-handling changes.
    e. Leadership of, or participation in, the worker level improvement activities, staff projects, or task force projects that affect their area.
2. Revise or develop their department to balance operations as much as possible when requirements of the schedule change. They may be assisted by staff:
    a. Establishing cycle times for all parts according to schedule.
    b. Establishing the sequence of operations (routing) within the department.
    c. Setting the points at which quality checking is necessary (hopefully done automatically).
    d. Adjusting cycle times and determining the number and sequence of daily setups so as to produce according to schedule while allowing for the current condition and capability of the equipment and people.
    e. Determining (or confirming) how much work in process will be required within each work center. (Production cards needed.)
    f. Determining or reviewing (with staff and higher managers) how much inventory will be required between work centers. (How many move cards will be required.)
3. Take whatever action is necessary to stay on schedule, or to have parts ready when others need them without exceeding work-in-process inventory limits.
4. Develop the capability of those who work for them.
    a. On-the-job instruction.
    b. Encouraging workers to enter various kinds of training programs.
    c. Cross-training people.
    d. Stimulating worker efforts to contribute to production improvement.

EXHIBIT 11–1 *(continued)*

Production engineers:

The tasks of the three categories of engineers below overlap. Companies classify the work of different engineers in different ways. However, if a concentrated plan for stockless production is implemented, clearly all engineers must spend time on the floor of the plant developing their plans into practice by perfecting them through first-hand observation and correction.

1. Quality engineers:
   a. Processing capability studies and machine studies; also assuring that processes are capable of meeting the specifications of the product, or, vise versa, that product specifications are not capriciously issued without regard to process capability.
   b. Developing inspection criteria: what constitutes a defect.
   c. Developing inspection methods with as much automatic inspection as possible.
   d. Monitoring progress on reducing defect rates; developing methods to establish the causes of defects and instructing others in the methods of establishing assignable cause.
   e. Attacking major causes of high-defect production.
   f. Working with suppliers:
      (1) First, to certify their quality and eliminate incoming inspection.
      (2) Second, to refine the quality of production processes used for incoming material to refine quality of parts.
   g. Developing or improving training programs on quality control.
2. Manufacturing engineers:
   a. Equipment selection and modification for stockless production needs.
   b. Development and improvement of tooling; tooling systems.
   c. Developing improved methods to transport material.
   d. Where possible, developing methods to move equipment for flexible layout.
   e. Setup time reduction programs.
3. Industrial engineers:
   a. Concentrating on methods studies, not on setting standards. (In fact, it may be a good idea to rename the industrial engineers as methods engineers to emphasize this.)
   b. Revising the layout of the plants and the various departments.
   c. Revising the material handling systems:
      (1) For material in production.
      (2) For material received from suppliers; receiving operations.
      (3) For material shipped.
   d. Establishing the sizes and types of containers used for ease of handling and prevention of damage as well as promotion of fast material flow.
   e. Studying the entire process of production for improving quality, speeding material flow, and eliminating unnecessary steps.

Purchasing agents:

A purchasing agent's job will be most different for those suppliers who supply parts which are made for the company, not commodities such as wire, hose, fasteners, and so forth.

1. Develop long-term partnership arrangements for a compact network of suppliers.
2. Develop a quality program with suppliers:
   a. For certification of parts they ship.
   b. For assuring that they have the correct specifications.
   c. For programs to improve the quality of both parties' production processes.
3. Develop a method to keep suppliers immediately and accurately informed of the role they play in the company's stockless program.
4. Develop a program to have suppliers ship as needed. Decrease the pipeline

EXHIBIT 11–1 *(continued)*

inventory from the supplier's last operation to the first operation where the part is used, but only as processes allow.

5. Revise the methods of negotiation with suppliers to reflect the requirements imposed by stockless production.
6. Be sure that adequate avenues of communication are open between the company and its suppliers, and train personnel in Receiving, Engineering, or elsewhere who have need to contact suppliers.

Production planning and control:

This job will change as the company moves into stockless production. It will evolve in anticipation of each new stage of development.

1. Start by decreasing the inventory while retaining the old system. Inventory cuts must be done to uncover problems and to better see how to make physical changes on the floor, so this must be coordinated with those who are resolving the floor problems.
2. Develop a system to plan production from the final assembly schedule (similar to the method described in Chapter 4). The planning method must evolve before conversion to a pull system, or concurrent with it.
3. Level the final assembly schedule as much as possible, a step at a time if that is what is required. Again, this must be done in conjunction with Marketing as well as Production. It guides a physical transformation process.
4. Develop a method to more closely link fabrication schedule to final assembly schedules. This can be done at first by basing the master production schedule more closely on the final assembly schedule.
5. Develop methods to estimate work in process at all points. This requires gathering information about why WIP is needed, and a method to generate the approximate number of cards or develop the alternative method for setting a limit on WIP.
6. Revise the schedules sent to suppliers. Start by stabilizing the supplier schedules, followed by the program to draw material from them as capability to do that evolves.
7. Develop a program to systematically remove production material from the inventory system as it currently exists and transfer the control of it entirely to the plant floor, or to whatever system replaces the stockroom.
8. A few bulk items, spares, and the like remain in the storerooms, but the number of inventory control personnel will decrease over time. They must be prepared to assume new jobs, possibly transferred to production control on the floor.
9. Develop methods to improve the handling of engineering changes and model start-ups.
10. Develop a new method for handling the bill of material so that Production Control maintains the bill. A new relationship to do this must usually be worked out with Design Engineering—one that does not allow the design of the products to be compromised.

This is a very brief summary of a very large conversion program. The work in production planning and control will be greatly changed. The rate at which the changes take place can only be determined by the rate at which the production processes themselves undergo change, and also by how great a revolution from the current system is necessary.

Data processing:

There is a fear that a stockless production program will eliminate the need for data processing expertise. In fact, many new programs will be needed: for leveling schedules, for estimating the work in process, for recording a flow of transactions, and so on. The result should be a decrease in overall systems complexity, but the implementation will strain the ability of data processing to generate programs for the purpose.

EXHIBIT 11–1 *(concluded)*

Accounting:

If the company plans to shift away from the use of work orders on which a cost collection system has been based, this is a huge demand on Accounting, almost like starting the cost-accounting system over. Along with everyone else, Accounting needs to be ready to rethink the fundamentals of what they do, so the change in type of work will be very great.

1. Shift from accounting based on standards, to accounting based on targets (more a recognition of a faster rate of cost change than a methodological change).
2. Collect costs by departments and processes.
3. Develop the means to accumulate the result of many transactions reported. For example, suppliers will not be paid for each shipment which arrives, but on a cumulation of shipments.
4. As lead times shorten, the entire process of accounting for the amount of labor cost stored in parts inventory begins to disappear. The labor cost is added at the time production is complete.

As with production planning and control, this is but a thumbnail sketch of a very large change in internal accounting practices. To accomplish them, the accountants obviously need to be prepared to change the accounting system to measure the new reality.

Exhibit 11–2 is a somewhat speculative summary schedule for implementing stockless production in a medium-sized manufacturing company which is not afflicted with serious problems to be corrected at the outset. That is, it is assumed that cash flow is not at a disastrously low level, the work force is not rebellious, most of the products retain a competitive niche somewhere in the marketplace, and so on. Of course, any exact sequence of activity depends on the individual case of any company undertaking the program.

ORGANIZATIONAL CHANGES

Study

The main objective of studying is to develop a vision in all key personnel of what the company would look like with a pull system in place. That may be either hard or easy to visualize, but, in any case, the number of changes to make are so great that they cannot *all* be anticipated in great detail. Therefore, study is necessary to be sure that everyone will be pulling in the same direction. To do that, everyone must have a basic and correct understanding of stockless production so as not to waste time doing things that later will be seen to have been incorrect.

Vision requires enough knowledge to anticipate the scope of the change and how most patterns of daily thought in manufacturing will be changed. Stockless production is only learned in detail by doing it.

EXHIBIT 11–2 Hypothetical summary schedule for implementing stockless production

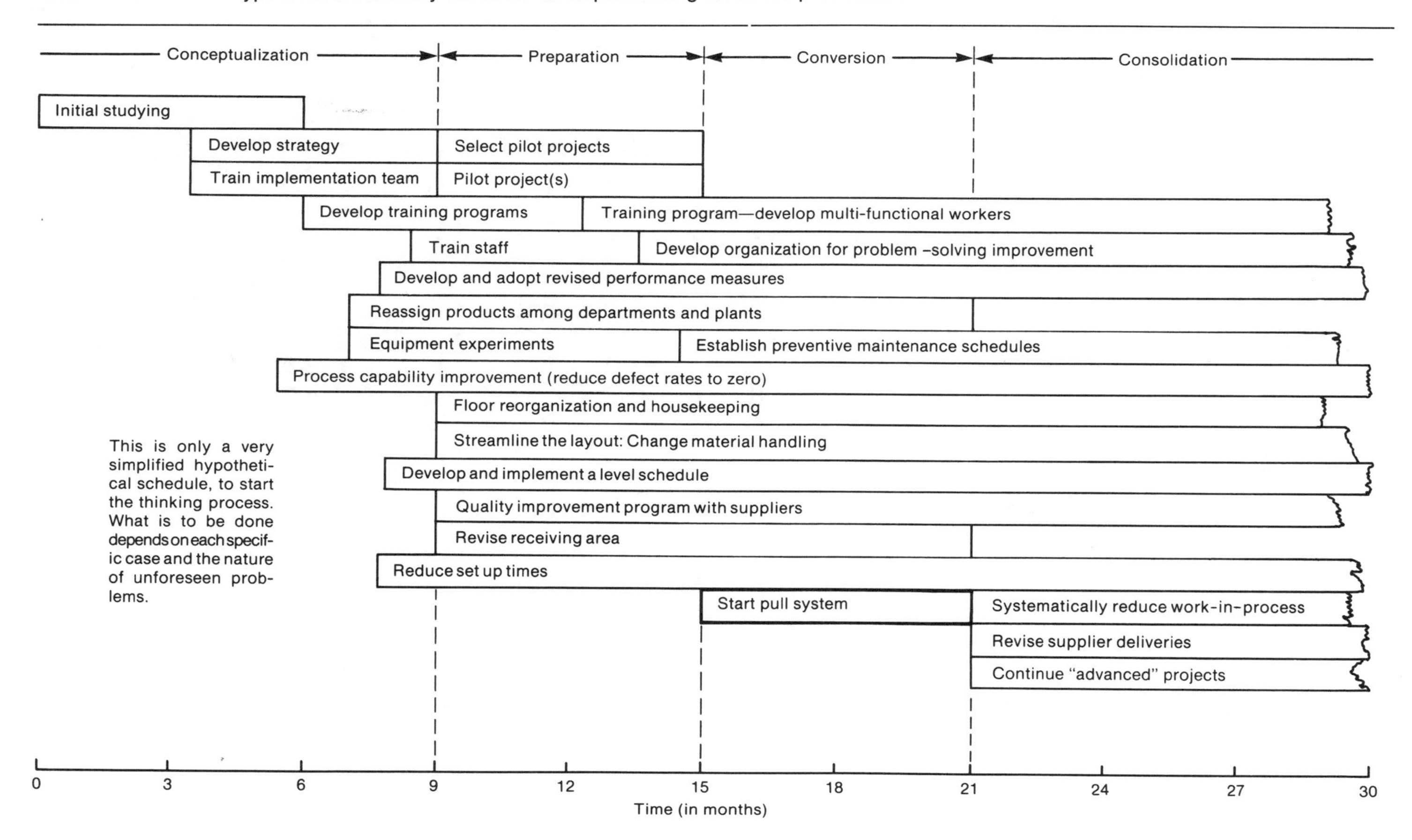

1. Top-management education. The top management must understand enough about stockless production to know that everything in the company will be affected. They must be able to truly lead. If they merely give assent to a program whose scope they have not grasped, they will soon be lost in the corporate conflicts that will ensue. Everything from marketing analysis to product design to cost accounting will undergo some kind of change, and the top management must understand enough to recognize progress as it is being made and give support to the correct actions. Only top management can devise and execute a change in company strategy, and they cannot expect success if they allow contradictory policies in personnel, accounting, marketing, finance, engineering, production, and materials management.

During the period in which major physical changes are taking place, the top management needs to visit the plant regularly, commend achievement, show concern, and make it known that they support the implementation team. They also need to know enough to understand if the implementation team is doing a good job. This role cannot be played unless the top management knows a great deal about stockless production and is prepared to see it make far-reaching changes in the company.

Who is top management? Several Japanese companies started their major activity by making presentations to the board of directors and getting their approval. If the board of directors is to know about it in general, the chief executive officer should know even more. In the absence of basic knowledge, the top management can insist, for example, that a pull system be instituted at a division which makes heavy generators: irregular order pattern, substantial custom engineering, project management, and some very difficult manufacturing processes. (That has been suggested, but is fraught with difficulties, beginning with how to establish a level schedule.)

During implementation, a key member of top management acts as the champion of the stockless program. That has been true at all the companies known to be successful with it, beginning with Ohno at Toyota Motor Company. It appears that a key member of top management must be prepared to assume that role. It is not necessary that top managers know every detail about how to make a process capability study or reduce adjustment during a setup, but they must understand what is generally occurring and be capable of explaining it at the highest levels of the company. The top management champions of Japanese stockless production companies have seemed capable of explaining a great deal of detail at the plant floor level.

2. Implementation team education. The implementation team consists of those persons responsible for making the major changes at the plant floor level and within other detailed operations. A team for converting a plant will consist of 5 to 15 persons of diverse specialities, including the supervisors of those areas in which activity is concentrated. The types of specialties which should be represented on the team depend on the existing conditions of the plant and the diversity of background experience in the persons selected. Suggested specialties to represent on the team:

Production planning and control. (Big changes here.)

Manufacturing engineering. (A major activity.)

Quality engineering. (Very important if defect levels are too high at the outset; not a major activity if defect levels are low. Something more than inspection experience is needed.)

Foremen. (Key persons. A foreman must coordinate the changes in his area so that production somehow gets accomplished while the changes are creating chaos during critical junctures.)

The specialties listed above are expected to have the greatest amount of floor work in an American conversion. If the company is primarily an assembling company, however, purchasing managers will play a key role earlier than they otherwise would. Other specialties include

Accounting.

Maintenance.

Purchasing.

Traffic.

Design engineering.

The implementation team needs to have the most detailed preparation, for they are the ones who must lead the plant floor through the confusion of the change. Reading about it and thinking about it are very helpful, but there is a very heavy element of see-and-do in stockless production. Knowing the concept is very different from seeing what to do when it is not working properly on the shop floor. Therefore some of the education has to be in terms of plant floor experience in the facility which is to be converted, and, in addition, it is very helpful to see the system in operation elsewhere. The preferred method of education in Japan is to give implementation team members several weeks of on-the-job training in a plant already successful with stockless production, but this may not be feasible in the pioneer conversions in the United States.

This kind of education takes time because it is necessary to build confidence, not just accumulate facts. People have to learn to adjust the pattern of thinking by which they accomplish their daily work, and that is a matter of developing skill. Therefore it is a good idea not to start before people are ready to accept skill building. No scientific evidence exists, but the author's observation of Americans learning about stockless production is that it takes three to six months to pass from the stage of skepticism and befuddlement to a point where confidence begins to rise.

During the educational phase for the implementation team, these people will have to spend some time learning and planning what they are to do. The time free of other responsibilities is necessary. This is also true of supervisors. (The Kawasaki plant at Lincoln, Nebraska, assigned an assistant to each line supervisor to absorb some of the daily hassle and give them time to think.)

The requirements for learning something substantially different and sometimes being imaginative about it are very important in selecting persons for the implementation team. The members of the team need to both have some experience and be receptive to new thinking. They also need stamina for the physical demands of floor work during the conversion effort. A few Japanese companies report reassigning personnel in preparation for stockless production. The implementation team members were their brightest persons in the age range of 35–45—old enough to have some experience but young enough to withstand the physical demands of the task. The education effort is really preparation for a shop floor leadership exercise of a lifetime.

3. Middle management and staff education. While everyone cannot be on the implementation team, there is a role for everyone. If everyone on the staff has a general concept of what is about to disrupt their life, it is very helpful, because most of them will be involved in projects to change how the plant operates. To take just one example, the system of data collection by accounting will probably change, which means that the accounting department has to have a project team involved in developing these changes and an accounting staff ready to implement them. If they have no understanding of why changes are taking place, a great deal of management time will be expended in combating resistance.

In some plants, a substantial reduction in inventory may induce a revised method for adding the labor cost to items in inventory, and very likely some plants will be changing from a job-costing system to a system which is more like process costing. The accountants are not going to do that without getting themselves mentally ready for a very big challenge. In all likelihood, almost every cost report prepared will have some changes in the method by which it is developed.

Therefore, the accountants need to be prepared for some hard thinking about how their job is to be accomplished. They cannot do that very well if they are unable to anticipate the kinds of changes that will be taking place. They will need to do some reading to prepare for it.

The accounting department is only one staff area. Each one will have to react to a big change. No literature detailing what to do in each special area is available, so the education must presently consist of reading about the system, receiving some instruction in it, and trying to determine how it will change their work.

4. Worker education. Nothing works on a plant floor if the work force opposes it. In stockless production, the worker has to take an active role in making improvement. The detailed organization of each workplace cannot be accomplished without the worker accepting this responsibility. Most people cannot accept what they do not understand, so some form of preliminary education is useful for the workers. It does not have to be detailed in the beginning.

In the American plants where a project has been well organized and the workforce basically informed, the attitude of the worker has not been a major problem. Most of them like the idea of accepting a little more responsibility for their own output, and the notions of stopping the process to correct problems is not one which is provocative. That is true of all the concepts of stockless production provided the work force does not receive misinformation about them.

Most of the Japanese stockless production plants have brief manuals for their workers, outlining what the system is, what their role is, and what the general rules are or will be. The manuals typically describe enough of the whys for instituting stockless production to satisfy questions of a general nature. In Japanese plants, foremen are responsible for more informal training of workers than is usually true in American factories. The American version of keeping them informed is a training program followed by frequent information about what is happening. Basic good sense is about all the guideline currently available for training the American worker, and so far that has not appeared to be one of the problems of implementation.

5. Discarding practices that block progress. During the period when interest in stockless production is rising, almost everyone in a company can agree that there was a lot of "sinfulness" in past practices even if they cannot agree on exactly what constitutes sin. One way to get past this period is to start changing to the forward look in performance measures, somewhat like the performance measures in Chapter 10. That takes some thought, so it will not be done overnight, but it stops holding staff and supervisors to performance measures which make it difficult to change the status quo.

For a great many American companies, two other actions seem to be widely applicable:

1. Get staff people closer to the production floor. That builds teamwork, but it does not happen by changing culture as quickly as it might happen with some reorganization. If staff teams work directly at department level with those who have to make the changes occur, there is little time for argument over the turf boundaries of various staffs.
2. Stop accepting excuses. These come in many guises: other people have not solved their parts of the problem; the equipment is not suitable for it; we do not really have the authority, and so on. Most of the problems can be summarized as not aggressively working on problems as a team. A problem to one is a problem to all, and the inventory can cover not only a great many physical problems but the comfort of making excuses as well. It does take wisdom to distinguish real problems from excuses, and management resolve to stamp out the known excuses as well. Only the top management at any location can stop the cycle of excuses by refusing to listen to them.

6. Developing the organization to solve problems. If the inventory is decreased without people being mentally and emotionally ready to overcome the problems that were the cause of the inventory, stockless production will not come to anything. A part of this is, to be sure, everyone understanding that the program is long-term, something much more than quick changes in the press department and card systems, substantive improvements made with imagination, not just the creation of appearances and a drive toward ultimate automation.

How to set up a problem-solving organization depends somewhat on the nature of the plant and the strategy of implementation. Some plants may be designed for stockless production, in which case the design staff anticipates as many of the problems of stockless production as possible, but it cannot foresee all of them. If stockless production is introduced into an operational plant, a process of evolution begins. In either case, the system develops and matures from the practice of it.

As the plant starts into the program, plans for future action come from the results of current action, so the implementation team must alternately teach and listen. The plant operates as if it were permanently in the start-up mode. Existing methods may be challenged and changed even if they were recently installed, and the plant is run as if it were a production laboratory, a point of view very helpful in developing the problem-solving organization. This organization needs to stimulate the following at the plant floor level:

Retaining stability in personal relationships while bringing different people together to work on problems.

Testing ideas, observing results, generating new ideas.

Identifying and correcting the root sources of problems through the promotion of causal analysis.

Stimulating suggestions and following up on them.

Japanese plants in a stockless production program swear by quality control circles (or small group improvement activities, or whatever name is given to them.) It is the method which works very well in that culture and which may work well in some American plants. However, there are many organizational variants for accomplishing this task: project teams, task forces, suggestion systems. Americans have a great variety of managerial styles and methods. The precise organizational method is less important than the achievement of plant floor level leadership and inspiration by whatever methods work in the local context. This is not to say that any method of organization will work, but only that there is no single style of leadership by which a program can be successfully developed.

Visibility in the plant is essential. The objective is to get workers so that they know what to do automatically, with a minimum of communication. All these plants feature methods of making problems visible. The style of leadership should not contradict that. The objective of the leadership is to develop the people to be capable of operating the plant themselves in a new way. Once the visibility systems are in place, much of the floor control is in the hands of the workers who respond to the various signals, and much of the correction of routine problems is accomplished the same way.

Moralizing about what to stop doing and how to improve is of some value, but somewhat limited. It is more useful to visualize how jobs will likely change for a variety of positions in the plant, so that people can prepare themselves for the change. The program to transform from the current condition into the desired one is a function of the changes needed at each plant.

Exhibit 11–1 shows some anticipated changes in various jobs as a result of stockless production. It gives an impression of the magnitude of the complete transformation based on an hypothesized average American case, but it presents very little detail. It has a few ideas for the impact of stockless production on the work of several major categories of work in a production plant, but it does not mention several important areas, such as maintenance and the tool-and-die department.

Everyone's job will be greatly changed, and the changes cannot be totally anticipated. The way to start is by developing a preliminary plan without a mass of detail, and then just start. If people are prepared to cope with both the anticipated problems and a few unanticipated ones, stockless production will materialize. All the companies which have gone into this have spoken about a process of discovery rather than following an extremely detailed plan—a little like having a road map, but not knowing all the detours.

Offer a guarantee of employment?

In the turmoil precipitated by stockless production, sooner or later many employees are going to ask about the security of their employment. If they contribute wholeheartedly to improving productivity, and then sales do not increase, what happens? Have they contributed to their own unemployment?

This is a very difficult policy decision. Any job guarantee given has to be subject to the fortunes of the company itself. In Japan, a core group of employees is considered to be guaranteed employment, but this does not extend to all employees. This agreement is usually not written, but understood. If the company itself has such a large reduction in sales that the guarantee cannot be honored, the management assists them in finding a new job. (When shipbuilding volumes dropped by 40–50 percent in the mid-1970s, the shipyard companies made agreements for their employees to enter other businesses, which were growing. Few of them missed any work.)

In the United States, this question is usually resolved by a seniority policy. The lowest seniority employees must leave first. Each company has its own version of some policy in this matter.

However, the question has not seemed to be one of the most difficult ones facing companies entering stockless production. People on the floor doing the work sense that if the company improves, the chances of its remaining healthy improve, and they have a better chance of employment with a healthy company than with one in trouble.

Unions are just beginning to grapple with questions of employment policy in case of displacement by automation. This issue has been known for many years, but it has never materialized into a serious immediate threat. Most of the serious issues have concerned the responsibility of companies closing plants or facing extinction. Although Japanese companies adopted stockless production from a fear of failure if they did not, all of them have been able to take care of any decrease in employment through attrition. Most of them went from a serious business condition to one of such growth that employment actually increased. The growth offset increased productivity of 100 percent or more.

Management who have considered this have generally concluded that they cannot truthfully guarantee their employees immunity from the unknown future economic fortunes of the company, so the best policy is the truth itself. The company will try not to automate employees out of a job and will only lay off if decreased sales make it a necessity. The most secure employees know who they are.

This issue goes both ways. There is an ever-present chance that the key persons on an implementation team will be attracted by employment elsewhere (in the United States). This is especially true if companies begin to recruit people with experience in this new system. In particularly critical situations, companies may wish to sign employees to a long-term employment pledge covering the expected term of the project, but even that is no guarantee. The most successful American companies (with or without stockless production) have had the policy of maintaining a strong training program and substantial talent to replace people who might leave. (Consider how many people are "graduates" of IBM or other leading companies.)

PHYSICAL ACTION PROGRAM

The plan for physical action takes place during the conceptualization phase of the project. Companies may do some experimenting as they develop expertise among the implementation team to solidify ideas on how to proceed. Every company has ongoing improvement programs of various kinds that are in progress when stockless production preparation begins, and one of the difficult organizational tasks is determining how to redirect these so that they fit into the total program.

Because of the multifaceted nature of the program as an entire package, it will probably take the appointment of area coordinators to keep the different improvement activities from trampling each other. (This has typically been the role of line foremen or supervisors themselves.)

A separate budget is possible for modifying equipment or for moving equipment and for implementation team members assigned to particular areas. However, much of the activity involves people doing direct production work so that in the end, costs and benefits as they materialize have been so embedded into the budgets and actual costs of the departments involved that an overall project cost is not cleanly definable.

The preparation stage consists of a large number of activities all occurring simultaneously: housekeeping, revising layout, selection of equipment and flow-path clarification, setup time reduction, improvement of process capability, improvement of quality control procedures, training people, and so on. How long it lasts depends on the rapidity with which progress is made, and that is hard to forecast very exactly. Those companies who started with people familiar with stockless production have not taken less than six months for preparation, and one to two years seems most probable when everyone is in a learn-as-you-go status.

The conversion to a pull system is the critical stage. It tests whether preparation has been sufficient, and it is the period of most intense work.

The planning for much of the overall development conceptually works backward from final assembly. The development of fabrication areas can only proceed so far without making assumptions about the kind of final assembly process they will feed. It is also true that Final Assembly cannot maintain a level schedule if the fabrication areas cannot support it and thus starve it out of material.

Pilot projects

Selection of a department or area as a pilot project is a very important decision. Sometimes more than one pilot project is wanted, depending on the purposes they serve.

Sometimes a demonstration project is desired, and its selection depends on the nature of support for stockless production in the beginning. A demonstration project would probably be in one of the easier areas so as to provide a quick victory to boost morale and gain support; but the logic does not always work that way, and some demonstration projects have concentrated on very difficult areas so as to be even more convincing.

From the point of view of a complete plant or company program, the pilot project needs to be representative of all the different kinds of problems to be faced. The project in the pilot area can run several months ahead of the rest of the program and serve for experimentation and as a learning center for everything—including cost systems. It is also a good idea to select an area whose operations can be isolated from the rest of the plant—not have a lot of cross-flows of parts with it. As the pilot project changes procedures and moves into a pull system, it will operate independently. It is very confusing to try to operate by two different systems at once, and this situation will also give an unclear picture of what will finally result.

In order to have a complete project, the pilot area needs to consist of a final assembly process and several work centers which feed it. Ability to improve fabrication areas is limited unless they feed a final assembly process which is running on a level schedule to which they can be balanced.

It is also desirable to have enthusiastic and aggressive people working the pilot area. The implementation team starts its work there, using that area as a place to try ideas and as a place for instructing the rest of the plant what to do. Some people understand the system intellectually, but cannot accept that it applies to *their* unique problems until they see it in action. The people in the pilot area ideally are those who can break ground first. The sequence of actions to take in the pilot projects is about the same as in the complete plant, but with a lead time of several months ahead of the complete plant.

Organization, layout, and housekeeping

Without making any radical changes in the current production-planning and control system, the first stage is to organize the production area so as to eliminate confusion and make the problems clear. This is very important and it may initially be very difficult. It is one thing to visualize how the system might work if a lot of things were different; it is another to systematically remove the obstacles to clear thinking which exist on the plant floor, so as to determine how to convert what really exists to what is possible.

This begins in a very humble way, with the organization and housekeeping of each work station in the area according to a version of the housekeeping and work rules:

Remove the trash. Remove everything not needed from the area.

Designate a location for all tools and materials, and keep everything where it is supposed to be.

Do not allow the cleanup to regress by allowing unnecessary items to accumulate in the area.

Remove the unnecessary work-in-process inventory from the area.

Simplify the movement of material as much as possible.

As this proceeds, introduce some of the regular quality-checking procedures, regular checks for preventive maintenance, and a setup time reduction program. This helps the workers keep in mind why the housekeeping rules are important.

Equipment selection may not be so easy. If setups are to be done rapidly, then one must designate precisely what parts are to be set up on what equipment; which means a study of what goes on what machine currently, with some thought given to that. If too many parts go on one machine, setups become difficult. One has to plan for a specific pattern of setup changes in modifying each piece of equipment and in developing the layout for it. The entire panoply of problems associated with setup reduction and process capability improvement are entangled with the decisions on what equipment to assign to what tasks and where to put it.

Streamlining the material and tool flow is a part of this. One of the most common and troubling problems is that a work center produces parts that flow into a specific assembly area, but it also produces parts for other purposes: spare parts, parts for other plants, parts for products that are demanded irregularly, and various other "emergencies." Many of these problems seem insurmountable at first, but as the easy problems are overcome, the more difficult ones also begin to resolve themselves. After a time, the workplace organization starts to approach the ideal of one part being found in only one place. An excellent way to think of it is as dirty flow being changed to clean flow, so that additional problems can be seen. Using the river of inventory analogy, many of the problem rocks can be seen under the surface if the flow is clean, so they can be attacked before they become totally exposed. Inventory is reduced whenever possible, using whatever methods are available through the production system now in use, and that keeps bringing the deep rocks a little closer.

This process is one of constant attention to operational detail. Exactly where to position a box of fasteners or a practical way to prevent chips and oil from splattering on the floor do not seem to be very momentous steps in the march to such a momentous change in production philosophy, but they are a vital part of it—the basic movement on which all the rest is built.

The implementation team has to spend a lot of time on the plant floor, and everyone has to reinforce the eventual goal to which the program is headed. Just by cleaning up the workplace, so many improvements are sometimes made that

People start to think they have achieved the goals when they are still in the early stage of preparation.

It is easy to be distracted and head in a direction with less potential.

At this point the implementation team has begun a program of daily action that will last a long time—probably one or two years, at the shortest. An example of this kind of activity was observed at American Koyo Company. The team meets every morning in a small conference room. They briefly review what was accomplished yesterday and what is needed today. A list of 15–20 projects is handwritten on the wall, with brief notations of status. Each day one of the projects may be crossed off and another added. Some of the meetings are long. Many are short and held standing up. When the meeting is done, the team moves into action. The next day they do it again.

How difficult this is depends on how tangled the flow of material is when the program begins. A potentially repetitive plant which has been operated as a job shop with much specialized equipment and long cross-flows of parts has a long, difficult task, and one that cannot be resolved quickly. They have embarked on a completely different strategy of equipment selection and material-handling methods. A lengthy period of detailed work at the very heart of the operation awaits them, and rushing the program probably means some very costly mistakes.

The expense and risk are in changing equipment strategies. To the extent that quality can be improved, setup times reduced and maintenance improved without major equipment selection expense, the program will have great benefit even if the plant never gets to a pull system and beyond. However, it can only progress so far without developing a smooth flow of material, so these are the critical decisions. At this point, it is good to remember the admonition of the pioneer companies in this: Gain experience with modifications of equipment you now have before plunging into major expense. The pilot project is very valuable for learning to read the rocks.

Improving process capability and quality-control procedures

One of the major problems which most American plants will face that Japanese conversions did not is defect levels. Most of the Japanese plants preceded stockless production with intensive campaigns to improve quality (reduce defect levels), so that they had only to refine their practices from a good base of prior development.

The idea is to eliminate the causes of defects which are easiest to correct first. Some sources of defects are very stubborn, so the material flow may be slowed for a long time because of some of them. For example, reducing the defect levels from painting processes typically is so difficult that a pull system has to deviate and work by keeping a little safety stock after these operations. If such a thing must be done, what is desired at a minimum is to not have so much stock that it cannot be held on the plant floor itself.

How low must the defect levels be in order to start a pull system? That depends, of course, on the nature of the process. If a process regularly creates exactly 20 percent defects, a pull system can be designed around it. But this is seldom the case; the defect rate is an average, meaning that the process goes totally out of control from time to time, and defects come in streaks.

Defects need to be sorted out before they go further in a pull process, and that takes time, but it can be done if failsafe methods are adopted by the work center. The real question, then, is about how low should the defect level be even if there is a sorting process? A rule of thumb is that it should be below 1 percent generally, and almost never above 3 percent, and then enough should be known about the process that when defect levels begin to rise, the corrective actions to take are known and can be immediately applied. The rule of thumb does not replace good judgment about the history of the process with respect to the demands which will be placed on it by the pull system.

Naturally, the goal is to drive defect rates all the way to zero with the pull system in place. You do not start with very many processes at the zero level. However, the program to improve process capability, reduce setup times, control tool use, and improve preventive maintenance should reduce defect rates below 1 percent except where the conditions are hard to control (painting, moisture absorbing resins, and so forth), providing many activities are under way which contribute to improved quality:

Developing the maximum amount of immediate inspection and failsafe methods.

Stopping to correct defective production conditions as soon as they occur.

Working on the reduction of setup times. Especially the study of ways to eliminate adjustments contributes to improving the quality and reducing defects.

Making substitutions in equipment used to provide more reliability and flexibility.

A quality-improvement program is also the place to begin with suppliers. If they understand why it is being done, reducing the pipeline of stock back to them may be helpful in assisting with this, but the initial objective is to avoid sorting defects from them or holding safety stocks to cover for their bad shipments. A pull system can be started from a valve inventory coming from suppliers, but we wish to eliminate the nasty surprises in the inventory.

In this stage, it is necessary to bring the organizational methods of problem solving to their full effectiveness. It is a period of learning how to do these things as a regular program, not as some form of unusual activity. The small group improvement activities, task forces, and methods for obtaining and acting on suggestions should begin to develop. If these are started fresh, some delay will be experienced because it takes time for these floor-level activities to mature into effective problem-solving approaches. People have to acquire problem-solving experience.

Likewise in this stage, the implementation team must refine the operating methods of staff specialists for coordinating their work on projects which contribute to overall progress. The pilot project is very useful for testing the ways in which staff engineers and others need to operate to make a large number of changes. Getting the right working relationship between staff people and the workers may be one of the most important objectives of this phase, depending on the nature of how staff people operated previously. To make stockless production work well, the staff engineers and other specialists need to work directly with people on the floor, and some engineers are reluctant to learn how to listen. They become impatient while floor workers are still trying to develop themselves to observe problems and determine their own roles in the system.

Experience of the Japanese companies was that stockless production greatly boosted the effectiveness of their quality circles because it concentrated attention on what needed to be done. A similar effect can be expected in American plants, provided American workers are developed for it by whatever methods seem appropriate.

Setup-time reduction

Getting started on this seems to be a great bugaboo, so a positive attitude about it among those who must do the work is important. The ideas for reducing setup time are often simple ones, but there are mental blocks in getting the ideas.

In large companies, at least, a setup-time bulletin giving times, techniques, and naming people and places may be big help. As long as the objective is accepted, how-to publications have a long history of success, and they provide recognition to those who have made breakthroughs while transferring the technology far and wide.

Studying the setup time reductions forces a great many of the considerations to come together. Much initial reduction of fabrication setup times can be done without a clear knowledge of how material is to be fed to final assembly. One knows that in a press shop, for instance, the die exchange times must be short in order to follow any kind of level schedule calling for many parts.

However, this development cannot run very far ahead of the development of final assembly. The closer the processes are tied directly together, the less independent the design of them is. It becomes important to consider how far a fabrication process will be from the line or from the processes it feeds, the kinds of rates and parts mixes that will be necessary, and material-handling considerations—how the parts will be presented to the using process. All these affect layout, potential sequence of setups, material changes, and so on, all of which can affect the selection of equipment used for specific tasks.

In summary, start setup-time reductions in fabrication areas in which one is sure that further developments are not going to change the basic equipment used. Remember, it conceptually keys on final assembly.

Material-handling design

We would like the material-handling system all the way back to key on the presentation of components for attachment at assembly as much as possible. *As much as possible* is a weasel-worded phrase, of course, because weight, damage, dirt, and the nature of other processes are all overwhelming factors in choice of material-handling methods. However, the ideal is still to fabricate in the sequence desired by final assembly and maintain positive control and orientation of the part to the point of its installation. That is automated material handling.

Few companies can start much of the material-handling conversion at so advanced a state, but it should not be forgotten. Generally, the initial problem is container sizes and the modifications of material-handling methods to go with them. For many plants and companies, the desired objective throughout a long period of stockless production development is to have a container size which fits the maximum number of parts and which therefore calls for great commonality in material-handling design throughout much of the process. Again, the advice of the pioneers is not to hurry this decision, but first gain some experience with different containers and equipment. One of the objectives of the pilot project is to experiment with this.

Leveling the schedule: Preparing for a pull system

Leveling the schedule in final assembly means revising the final assembly line or area to accomplish this. If a line is dedicated to only one model, this is elementary in concept, but if several models run on the same line, revisions are necessary to begin to reduce the length of assembly runs. Records are kept on how level the final assembly planned schedule is and of how closely the actual production adhered to that. It is not unusual to have planned schedules that are level but that suffer from material problems (or others) which cause one- and two-hour delays in changing the line. It takes time to revise the line stations to reduce the run sizes. Less material for one model must be on hand. Material for subsequent models must be positioned. The entire material-handling and material presentation system for final assembly must usually be reworked, but even where the lines run mixed models at the start of the program, there are usually problems which prevent the line from running a level output in practice.

How this starts depends on the proficiency of final assembly at the outset, but if it is not at all smooth, one starts by just trying to keep the output of the line linear to schedule. Can we assemble the same number in the second half of a shift as in the first half? If not, why not? Then work on making the number assembled each hour the same. Then decrease the run lengths in assembly (if it is going to be necessary to run mixed models) and try to keep a linear output with rapid changeovers.

All that will help build skill, but many of the delay problems in final assembly come from defects which arrive at the line. As the fabrication operations which feed final assembly improve their quality, those problems should decrease. As they do, it becomes feasible to stop the line, on the spot, to correct the problems which occur in assembly. In most cases, it is not possible when starting the program to have zero defects arriving at the line, so there will probably have to be some presorting activities to prevent excessive line stops. In short, it is not difficult to think through what needs to be done to arrive at a final assembly process which is linear in actual output, not just by plan. However, doing it has proved to be a major exercise. All the uncaught mistakes arrive at final assembly.

From the point of view of scheduling, development of a level final assembly schedule includes more than working orders and forecasts into a pattern of daily schedules. It must take into account the limitations of the production process being pulled by the schedule to handle model mixes and option mixes. Those companies that had never before tried to develop a mixed model final assembly schedule say that this is one of the hardest things to do. The schedule has to run first at final assembly before it can be transmitted to the rest of production. It appears to be a normal process of learning: do not start with anything too difficult, then increase the complexity of the assembly schedule, observing the capability of the line to run it.

At what point is the actual schedule level enough to install a pull system? Again, there is no clear answer. It is when a process appears capable of being linked and the material will flow smoothly, so try it and find out. Feeding processes are normally linked to the final assembly pull system one at a time for trial, but without removing the crutch of extra inventory.

Conversion: Installing the pull system

The pull system can be installed when a review of the material flow shows that containers, procedures, and material handling are in place and it is possible to estimate how many cards to issue. The pull system can use markers or signals other than cards, but, assuming that cards are issued, they are initially used just to test and refine the organization for them. The stockpoints and signal methods have to be installed and perfected. Then it is usually discovered that prior development was inadequate. Setup times need further reduction. Work center layouts need more thought. A process thought to have a tight quality-checking system does not, and so on. Every company has reported that this moment of truth has been difficult, not for every installation, but for many. The first step into the pull system requires much work just to get a complete operation so that it basically functions. Success comes when there is no longer a need to keep a little extra stock somewhere just in case, so that the stockroom or its equivalent is no longer needed.

While all this floor work is going on, the reporting systems are also undergoing conversion. If a production area has previously operated by job orders, the basis of the job orders is disappearing during this period. The accounting system changes to different methods of collecting data—cumulative counts. If successful, the normal transfers of material in and out of stockrooms disappear, so the accounting and materials reports which have been based on these transactions have to change, and these new forms of reporting have to be tried and tested.

Certainly one of the most revolutionary steps is the discontinuance of inventory transactions as a centerpiece of control. It is hard to make this change, and the transition period while a plant has one foot in the pull system camp and the other back in the push system camp is very confusing. It is for this reason that when a plant of any size and complexity begins a conversion to a pull system, the desire is to do it as quickly as possible. The two systems confuse each other. Again, the word from the pioneers is that once this conversion period begins, you have about six months before one or the other system has to collapse.

Although the amount of inventory in stockrooms is reduced during the preparation period, that which remains will be doled out to the floor for consumption at some time during conversion. This is an interruption in smooth flow, and it can obscure what is going on if not done carefully.

The implementation team works through the plant, area by area, guiding people in the conversion to the pull system. It is a period of intense activity and a lot of confusion, which can only last so long, like a conversion to any new system, but it is more difficult than most. The degree of difficulty depends, of course, on how different the pull system and the physical changes are from what went on before. A plant which has operated as a repetitive manufacturer for many years certainly has to make fewer changes in the patterns of daily thought than one which has operated as a job shop.

While the burden of leadership at the floor level falls on the implementation team, everyone else becomes familiar with stockless production during conversion, if not during preparation. All supervisors and staff take part in the conversion in some way. It is a difficult time, and one that cannot be accomplished unless everyone is prepared for it, both by training and emotionally. Several of the Japanese plants which made the change said that they spent several months building the morale of the entire workforce for the stress of the conversion period. The regular appearances of the top-management champion during the conversion are important in sustaining morale.

The conversion and of the preparation for it are a function of whether a plant is heavy in fabrication or whether it is mostly an assembly plant. An assembly plant still needs to go through the same preparation and conversion, but most of its feeding operations are from suppliers, so much of the long-term effort will be in working with suppliers. It is still true that the first effort is to improve quality from suppliers, not to establish a pull system with them until the basis for one is established in their own plants.

Most of the conversions to a pull system inside plants have concluded with some inventory left from suppliers being held in valve stock. Because of the transport system of Japanese industry, it is only to be expected that American companies will have higher levels of raw material valve stock than in Japan. It is the elimination of work-in-process material from stockrooms that is initially desired.

Consolidating the system

The end of the conversion period comes when a pull system is in place for most of the departments and processes. Operations that cannot function by a pull system are an exception and not the rule. The system may not work well, but it is basically in place and functioning, and the old system is gone.

By this time, most of the staff and supervisors should have enough experience with stockless production to begin to take advantage of that fact. The period of most intense leadership for the implementation team comes to an end, and the objectives of stockless production become diffused throughout the company. Many more people are now ready to continue the work of changing the company's methods of operation to make sure that stockless production is not lost but is used to make further progress.

Until a pull system is in place, it is difficult to refine the more advanced methods, such as development of U-lines and group-technology methods. This is a period of restudying how to improve operations in view of the added power which is provided by lower inventories, lead times, and defect levels. By going through a very careful review of methods, work center by work center, and working through the entire plant again and again, the company will greatly improve productivity, and here is where some of the major gains come.

This pattern of work can now continue with suppliers, and it can be done with much more confidence because the company's internal operations are ready to receive an improved level of performance from suppliers. More specifically, it should become clearer how to revise the receiving operations of the plant, and the schedules given suppliers are stabilized so that they can be expected to service a company which is equipped to provide help if they need it. The purchasing agents have a good grasp of what they are trying to accomplish, and they can develop (with the production planners) a scheduling system and delivery system that convert the suppliers into something much more like a network of work centers that happen to be at a distance from the plant. Until a reasonably level schedule is in use at the plant and it can be stabilized down through the fabrication operations to lower-level parts demand, the company can only talk to the suppliers about what it *plans* to do.

The preparation with suppliers involves schedule, specifications, defect levels, and material handling. Unless a customer company has a semblance of the physical system they want in place, they cannot tell a supplier what type of container is wanted or what the delivery system will be—how to come the closest to a direct hand-off of parts from the last operation of the suppliers to the first operation of the customer.

There is a great temptation for purchasing agents to press suppliers for too much too early. Once they can begin to work with suppliers as work centers outside the walls of the plant, they can begin to determine what mix of parts should go to what suppliers in order to provide each supplier's own production processes with a schedule for a stable overall volume, and how to communicate mix changes within that. By discussing the transfer of production from one physical facility to another, the purchasing agents can begin to develop a production network outside the plant. It becomes possible to institute immediate feedback on quality and other problems from the customer to the suppliers' points of production, and so on. The need is to reduce the number of suppliers to a group with whom communication can be established on stockless production. There are many considerations in deciding how many to have and the risks of having some as sole sources of supply, but if the suppliers themselves move into stockless production, the risks should be reduced.

The fears of supplier companies themselves should be allayed by recognizing that the volume of production for an entire industry should not decrease in total. What is desired is to simplify the flows of material within an industry. Only the suppliers who cannot deliver quality parts on schedule need fear.

From what has been observed in both Japan and the United States, the process of receiving will probably be revised something like this: The receiving area will have to process a higher volume of transactions at lower quantities each. More trucks will arrive, but bear in mind that the total volume of material received should not increase. The trucks are more likely to carry a mixed load of material, only part of which may be unloaded. Therefore the truck loading and the receiving system must quickly sort out what is wanted from the rest and get the truck on its way.

The problem in receiving is to process the inbound material into an order from which it can be drawn into the plant. As much material as possible will go, not to a stockroom, but remain in Receiving, which will act as a staging area for the plant. The principle here is to keep material in as nonrandom a fashion as possible, so the receiving area itself may become a superstockpoint. In order for this to be done, the pattern of arrival must be a regular one. Trucks and material cannot be processed in this way if material arrives all in the morning or in random bunches.

Inspection cannot remain a bottleneck. This is only possible if the inspection activity for incoming material is reduced, and if the pattern of receiving is made regular enough that inspection itself becomes a fast-flow activity that processes material as it arrives. (It is the queues in inspection which cause the major delays, just as is true in any job shop work center.)

The adventure of revising the system of deliveries in the United States can presently be dimly foreseen. It is too early to speculate much more about exactly what changes in transport methods and receiving methods can be developed. A company is well advised to work its way into the steps of stockless production, which can be projected from the experience of others, before getting involved in major efforts to revise transit pipelines, which are pioneering in the unknown.

Another area in which consolidation takes place is in the ancillary systems which surround production. For example, the method of starting new models must be established to function smoothly in stockless production. The grouping of engineering changes can be worked out. As the lead times and inventory shrink, these kinds of changes become possible.

The tooling and production support functions can be restudied in light of the advances made. Usually the work of tooling becomes more level, and, as a result of the work done to reduce setup times and improve production capability, the tooling shops can usually become more standardized in the way they operate. As described earlier, the Japanese plants on this system often have pull systems of control operating for the tool shop.

Not the least of the consolidation efforts is the conversion of stockless production into a people-control system on the plant floor. Systems of lights and signals of various kinds allow the plant floor to be controlled "automatically" by the work force. The Japanese plants on the system have typically had a great upsurge of improvement ideas from both staff and workers during this period. As people use the system, they become better at it. The more that is done, the more people see to do. The system of work-in-process control begins to be used to periodically decrease inventory in increments so that more and more detailed problems are uncovered and overcome.

One of the hazards during this stage is, again, that people will regale too much in the sense of accomplishment that a pull system is finally working and fail to pursue the use of it for further gains, and it is in this stage that the major improvements are made permanent and skills honed for making further advance. Without doing this, the company risks staying at the class C level in stockless production. The top management needs to be aware of the danger and be prepared to rally the company to overcome this.

At some rather vague point in this, consolidation becomes a matter of what Japanese companies have called tuning up the system. This means becoming adept in its use so as to employ it to make further advances. The whole concept is to first improve the operation, then improve the equipment and facilities. At some point in the consolidation stage, the operations are substantially improved, and the company can begin planning how to incorporate advanced technology such as robotics and continue the program on toward full automation. All the practices which have been put in place have prepared people to take these steps.

APPENDIX: TOKAI RIKA COMPANY IMPLEMENTS THEIR TOTAL PRODUCTION SYSTEM

Tokai Rika is located in Aichi Prefecture in central Japan, where there is a head office and three plants. Founded only 30 years ago, it is a major supplier of automotive parts to Japanese auto manufacturers, and they also count several foreign auto manufacturers among their customers. They produce seat belts, ignition switches, cigarette lighters, shift levers, turn signal combination switches, light switches, sensors, and fuse boxes—almost all the types of switches to be found in an automobile.

Tokai Rika has three plants. The largest is the Otowa plant, with about 34,740 square meters of floor space and 1,600 employees, located in the Mikawa District, a rural area. The story focuses on that plant although Tokai Rika introduced the total production system in all three plants at about the same time.

A great concern for their employees pervades the history of the Tokai Rika plants. The company is unusually strong in this, even in Japan, a country whose industry is generally noted for it. They believe in getting things done through the cooperative effort of all the people, and that has been a strength since the founding of the company.

The company began quality control circles in 1963 at the time that movement began. By 1973 they were implementing quality control circles at their subcontractors. However, work with the quality circles has not always gone smoothly.

Many of the people who work for both Tokai Rika and their suppliers are from farm families, part-time farmers and wives of farmers. Because of farm work and other obligations, they cannot give much of their time to quality circle meetings aside from the time on the job.

There is field work, and the women must see to their families. About 45 percent of the work force is female.

Because the women do not wish to work while their children are young, there are few women in the work force between the ages of 23 and 38. The average age of male employees is 28; the average age of female employees is 42. At the Otowa plant, they refer to the workers as mothers and their sons.

However, many problems are discussed at work. The Otowa plant set up very effective talking tables in all areas of the plant, near each workshop and along assembly lines. They sometimes call these the farmers' markets, because supervisors put materials on the tables for discussion. For example, a basket of defective lock cylinders will be placed on the tables with no comment. The unspoken question is, "What can we do to correct this?"

At times, the quality circle meetings took place in employees' homes after dark when the chores were done. The people even organized to help each other with farm work in season.

A great deal of effort is spent, not only in quality circles but also with suggestion systems and in other ways, to create and preserve a situation in which the workers react much as the boys in the story of Tom Sawyer painting the fence.

The workers should have the attitude, "Oh, so you want to decrease the flash in the lock cylinder moldings. Well, we know some things that might help with that. We might even be able to figure it out better than you can."

The implementation of the total production system at Tokai Rika is interesting because it illustrates the combination of strong leadership within an atmosphere of participative management. The management had to be both tough and gentle at the same time.

Three of the principal characters in this story are the following:

Yoichi Kato, managing director, Tokai Rika Company	Mr. Kato was the top-management standard bearer for the program. Though not part of the implementation team at the Otowa plant, he visited it frequently, two or three times a week. He viewed key plant floor improvements as they were made. He (and other top managers) made it constantly known that they supported the implementation effort. He cleared obstacles—discontinued traditional practices as it became apparent that they interfered with progress.

Tamae Okuhara, production control manager, Otowa plant	Mr. Okuhara was the man with floor experience in many areas of the Otowa plant, and a dominant personality by Japanese standards, what Americans might refer to as the bull of the woods, or the mover and shaker.
Susumu Okada, manager of administration, Otowa plant	Mr. Okada came to the Otowa plant from the head office. He was a member of the planning team for the conversion to stockless production. He was the resident expert on how such a system should work, the one with the problem-seeing eyes on the shop floor. He was formerly involved with implementation teams in Toyota plants.

Okuhara and Okada were only two of the floor generals of the implementation effort at the Otowa plant. Several other people were instructed and became part of the team, which began instructing everyone until finally everybody was part of the team.

THE OLD SYSTEM IN CRISIS

Prior to 1973, Tokai Rika was little interested in stockless production, Toyota's kanban system. They used a monthly ordering system for material control, and the plants operated by much the same methods as those used the world over at the time. By most standards, they did a good job. The employees were dedicated, and they had taken many measures to improve quality, many of them through the network of quality control circles.

Then came the Arab oil embargo. Lights went dim all over Japan, sales dropped, and schedules went into upheaval. Work-in-process inventory ballooned from 13 days to 24 days in only a few months, and that was the major reason for total inventory jumping from 24 days to 36.2 days on hand at the same time. What happened?

Demand decreased.

Model mix variations increased.

Inventory increased.

The shop floor got out of control, and there was panic expediting.

Shipping dates were missed.

The missed shipping dates brought an immediate negative reaction from Toyota, Tokai Rika's major customer who bought about 50 percent of total output. Toyota began to consult with them earnestly and frequently about this condition.

In early 1974, Tokai Rika worked very hard to improve the existing system. They tightened inventory-control procedures and improved data accuracy. They greatly improved the accuracy of production counts. The increased accuracy allowed them to cut safety stocks, which in turn allowed them to shorten lead times, and the shortened lead times allowed them to operate on the basis of better forecasts. They

1. Cut inventory in half—down to 17 days on hand of total inventory.
2. Reorganized to manage inventory by a system, not by the seat of the pants.

This effort stopped the missing of shipping dates, but it did not end Toyota's interest in the reliability of Tokai Rika operations. The lights were still dim and business conditions tough. After the oil embargo was lifted, it was still feared that another one might occur at any time, and Toyota feared that it might not be so easy to survive a second oil shock. Toyota insisted that Tokai Rika deliver to them by the pull signals using kanban cards. Tokai Rika's managers began to see that a kanban system might be useful for their own internal operations. Tokai Rika's top management studied the problems and discussed them among themselves for a long time. Finally they concluded that they were in absolute need of stockless production. President Immura (now chairman) called on Toyota's Vice President Ohno, who invented the kanban system, in order to request Toyota's support.

The first attempt: October 1974–September 1975

By the middle of 1974, Tokai Rika had become one of 25 companies in a Toyota study group on the subject. Tokai Rika managers visited Toyota. Toyota managers visited Tokai Rika. There was some internal dissension about it because they feared they might not be able to replicate what Toyota had done. Their conditions and people were different; they had unique products; they had customers, other than Toyota, who might not understand—not a pleasant prospect to jump into without due forethought. However, they had two strong motivations:

1. Energy was rationed and times were tough. All employees could see that at home and on the job.
2. Their major customer kept urging them to try it and promised a lot of help if they did.

Finally they decided to become an exemplary member of the Toyota supplier family by implementing it swiftly, perhaps even setting a record. They spent several further months in preparing themselves to do it the right way. Their initial goals were ambitious:

1. Cut inventory with the following target for the first year:

	Starting inventory March 31, 1975	*Inventory target March 31, 1976*
Finished goods, days on hand	9.1	4.5
Work-in-process, days on hand	8.0	5.0
Total days on hand	17.1	9.5

   (They were not greatly concerned with raw material stock at that time.)
2. Rationalize production (a British phrase meaning to improve productivity, streamline flow, and so forth).
3. Develop the ability to survive the tough business conditions which were expected to persist at that time.

The first two of these objectives were considered necessary to achieve the third—survival. They wanted to innovate with production methods in order to have a low break-even point. Originally, they wanted to break even at 80 percent of the 1975 sales level, and this was used as a selling point for both management and employees. In effect, if everybody helped to get the break-even point down, they helped to preserve the continuance of their own employment. That was the message. They began thusly:

March 17, 1975: Finish preparing a project-implementation schedule.

March 20, 1975: Complete preparation of operation manuals for everyone.

March 31, 1975: Take inventory.

April 1, 1975: Start changing the plant to stockless production.

In the ensuing three months, they held 19 major training meetings and 2 meetings to present improvements to management. They raced pell mell to get the card systems in place and rearrange the material flows to do that. By midsummer they knew that progress was not their most apparent product. It was time for soul-searching and reviews of what had gone amiss. Reasons for failure:

1. They did not work enough on the physical changes required, so:
   a. Setup times did not decrease much and lot sizes remained too big.
   b. The introduction of small standard containers was therefore stymied.
   c. They had to continue operating too much by gross-to-net calculations of the inventory. They could not get it out of the stockroom.
   d. They were really operating in many places by large lot order point methods.

2. They were reluctant to accept the chaos which would result from discontinuing the old system in the plant, and people found it confusing to try to operate by two systems at once.
3. They had a plan, but were emotionally unprepared to execute it when the magnitude of the changes became apparent. For example, they underestimated the need for multifunctional workers.
4. The implementation team had concentrated on material movement changes. The necessary equipment changes, layout changes and, operations balance corrections largely remained undone and, in some cases, unseen.
5. Most important, the implementation of this system was a kind of intense revolution which Tokai Rika's top management could hardly imagine. They had researched the actual status of the system in several companies that had implemented it before them. They thought they had anticipated the kinds of potential problems that could occur during implementation, and they had prepared for them. In spite of their careful preparation, they were not completely ready to deal with the great confusion that came over the plant.

Despite all these troubles, they did succeed in cutting the inventory a little bit. Most important, they revealed to themselves the kinds of issues they needed to deal with in order to make implementation a success. They decided to fall back, regroup, and renew the effort. They did not give up.

According to Professor Jinichiro Nakane, a consultant to Tokai Rika, the story of this failure is typical of most of the failures experienced by companies that have ventured into stockless production. The top management and the implementation team must gird themselves to manage a revolution, and they must prepare the workers to struggle through it.

The failure also occurred despite a very strong program of participatory management and well-established quality circles. In fact, in the early going, the workers have to be guided. Until they learn to recognize the evidence of problems provided to them by stockless production, they usually cannot contribute imaginatively. Some of their suggestions go in directions counter to that required by stockless production.

Successful conversion to a pull system: October 1975–March 1976

In the summer of 1975, the people of Tokai Rika restudied stockless production and made preparation for a second try beginning in October. They received help from Toyota. Masters of the art from Toyota toured the Otowa plant.

1. The top management studied the matter and resolved to support the implementation team visibly and consistently.
2. The implementation team studied how to identify wastes; that is, as the depth of the river of inventory was lowered to uncover the problem rocks, they developed methods to break up the rocks. They studied methods of setup-time reduction, workplace organization, tool management, and so on.
3. They began preparing both themselves and the work force emotionally to work through the confusion that would result from the conversion.
4. They selected two assembly areas and some feeding fabrication centers to be the pilot projects. From there, they planned to use these as demonstration areas and convert other plant areas by one revolutionary project at a time over several months.

When they began in October, they worked feverishly in the pilot project areas, but also began a plant-wide housekeeping campaign. This required very difficult work with the operators throughout the plant, but it was most important at this point. Area by area, they then worked through the plant in this fashion:

1. Reorganize the final assembly area and level the production schedule for it.
2. Go to the feeding fabrication work centers, decrease setup times and decrease lot sizes.
3. Follow up by reorganizing the layout for a visible flow of material. Introduce the cards just for housekeeping at first. Set up the stockpoints and get people accustomed to using them. Remove old containers from the floor.
4. Keep following the principle of cutting the work in process between operations. Use standard containers. Continue to cut setup times.
5. Introduce or intensify the preventive maintenance programs to reduce downtime. Set up record keeping and standard routines for caring for tools and dies.
6. Train operators to become multifunctional workers so as to be able to execute whatever level schedule is required.
7. Discontinue the use of the stockroom.

What differed from the first effort was that they persisted in making physical changes until parts for each product in every area of the plant could go by flow and be removed from the stockroom. They went area by area, reviewing and attacking problems step by step. They had meetings with staff, supervisors and workers day and night, stand-up action meetings and sit-down think-through meetings, and all of it dedicated to a very single-minded goal: get the inventory out of the stockroom.

Many meetings lasted all night. Many workdays were 18 hours long. Many members of the implementation team said that during the period from October 1975 to March 1976 they left the plant to go home only three or four times. (Implementation teams for other plants and other companies relate that their conversion effort was just as intense.)

By March 1976, the pilot projects were working well and the rest of the plant was on the way. Very little use was made of the stockroom for active parts. It was not working well, but the old system was gone and the new stockless production system was in place. Most of the success was achieved during the last five months of furious activity. The accomplishments:

1. Labor content was reduced by 31 percent during the year.
2. Inventory was down to 6.2 days on hand. (Note that they beat their original goal of 9.5 days on hand by a wide margin.)
3. Plant space was released for new activity. They had removed inventory space and closed up the distance between machines.
4. Improvements were made in production processes themselves. Process control had improved.
5. The thinking of production engineers and other staff was converted.
6. They improved their ability to react to changes in total volume and to changes in product mix. They became *more* flexible, not less.
7. Top management was fully converted—not a trace of doubt left.

Consolidating the system: April 1976–March 1977

This was a year of expanding the system, getting it to work well in the Otowa plant and extending it to the plant's suppliers. It was also a year of developing the workers in the use of the system. The most intense activity for the implementation team was over, and they trained others to give instruction in stockless production.

The quality control circles became small group-improvement activities. The scope of their work extended well beyond issues of quality. In fact, the rate of participation in the circles increased from the 70 percent range to a consistent 95 percent-and-over participation. Before stockless production, the worker group often had difficulty seeing through the confusion to understand what a problem really was. After the conversion, problems of every type became easier to see, and often easier to correct.

The programs to improve productivity continued unabated. In fact, the momentum increased as supervisors and workers understood that they needed to contribute improvements. True, many of the improvements are somewhat trivial, but the cumulation of them adds up to a big impact. For the workers during this period, the new system became *their* system. They had responsibility to make material flow through the plant. They had responsibility to respond to trouble points in production. Morale shot up by a measure commensurate with this newly discovered responsibility for making things happen. There is a great difference between operating a system and being operated on by one.

All this floor-level activity is stimulated by a program similar to management by objectives. Management presents targets for improvement, very general ones at the top level, more specific ones at lower levels. For example, top management may only set a target of a 10 percent cost reduction. One of the ways this could be translated at a departmental level is for a 20 percent reduction in tool breakage. That gives the small group-improvement teams a good direction for suggestions and projects. They know that suggestions about tools will get a very welcome reception, and the staff will be eager to assist them with ideas on that subject.

The foremen at the Otowa plant are all involved in the small groups, and they also deal with suggestions in more direct ways. The way in which their performance is rated stimulates them to do this, and the performance rating system evolved during this period. They receive weekly and monthly ratings by a point system that gives them positive and negative points in the following major categories:

Product and process improvements.

Quality of product (decrease in defect rates or suggestions for improved utility of the product in use.)

Cost reductions.

Morale of the people they supervise. (This is measured by attendance rates and the number of improvement proposals they make.)

Production to the schedule without trouble.

Safety.

This performance rating system emphasizes long-term accomplishment, not short-term numbers. Tokai Rika has a productivity-quality rally. It is a very important part of making stockless production a permanent establishment—a way of life in production.

During this year, Tokai Rika began instructing their suppliers in the use of the system, beginning by asking them to use the move card system for the delivery of material. Not all the suppliers were receptive, but many of them were, and representatives from Tokai Rika went from supplier to supplier, instructing them how to apply stockless production within their own plants. In this way, the Otowa plant began to decrease the raw material and purchased-parts inventory.

Several of the small suppliers are really subcontractors doing subassembly work using inventory on consignment from the Otowa plant.

Developing the system with these subcontractors was almost like extending the system to a remote part of the Otowa plant itself. Of course, extending the system to some suppliers, such as those for steel or zinc alloy, was not a feasible matter. However, the advent of stockless production did not increase the amount of inventory held by suppliers for Tokai Rika. Just the opposite occurred, and Tokai Rika worked hard to make sure that the suppliers' performance could keep up with them.

By cultivating the system during the year of consolidation, Tokai Rika management assured the company continued benefits for many years. By the end of the year, everybody was on stockless production, ready for further action.

Tuning up the system: April 1977–December 1977

Following the expansion of stockless production throughout the Otowa plant and all of its receptive suppliers, the people of Tokai Rika established methods within their total production system to maintain a steady pace of improvement. This is called tuning up the system. It is not clear precisely when this phase begins, nor when it ends. Perhaps, in a sense, this phase is still in progress at Tokai Rika in 1982.

Emphasis on this was very important for management. There was within Tokai Rika always a feeling that enough had been accomplished, and now perhaps everyone could rest a while—just let it run itself. However, management had to reinforce the total production system as a condition of permanent improvement. The business of the Otowa plant was to make improvement and become a production laboratory, not just a place to churn out products. This psychological change is vital.

The total number of solved themes from the small group-improvement teams increased tremendously—from 1 or 2 per year for each group up to 10 or 20 per year for each group. In some departments, the number of tools the workers built themselves went up 5 or 10 times. Because of the process control accompanied by immediate inspection and feedback, the number of complaints from customers decreased. Defect rates continued to drop down to hundredths of percents. In fact, the problems with defect rates became so trivial that attention in quality improvement came to focus more and more on the final function of the product and how to improve it with no added cost.

WHAT TO DO FOR AN ENCORE

The Otowa plant has continued to improve. Although sales increased and the complexity of the products increased, the inventory level has continued to decline slightly. Based on aggragate figures from the end-of-the-month count, the inventory levels of the Otowa plant in June 1982 were as follows:

Category	*Days on hand*
Raw material	2.5
Work-in-process	.75
Finished goods for Toyota	.30
Finished goods for all other customers	1.5
Total days on hand	5.05

This graph was derived from sales which appeared to be corrected for inflation. The number of employees include *all* employees, not just the production workers. 1973 = 100.

The market share of Tokai Rika has increased in most of the product lines in which they are in business. Their sales even increased in 1974, a year in which Japanese auto sales declined, and the year during which they had great internal operating problems. The improvement in productivity is shown by comparing the progression of sales per employee. What is most clear is that institution of the program continues to pay off, year after year.

Naturally there are a few voices of dissent. Some think that the improvement is being achieved by too much reliance on the increased facility of workers, and that not enough programmed automation is being installed fast enough. Increasing attention to this will probably be in the future of Tokai Rika. However, the motivation of the company is still to guarantee jobs, to maintain a low break-even point so as to survive a downturn without rapidly adopting methods that would displace people from potential jobs. They will evolve into full automation from their present position at a rate which depends on employment requirements and business conditions.

12

Stockless production and manufacturing strategy

Strategy is more than simple statements of general intent, or more than even such statements embellished by guidelines for action. It is a unifying direction for large numbers of people. Anyone can make a statement of purpose, but it takes much more than that for strategy to be valid. Strategy becomes a theme by which people at remote locations who may hardly know each other develop a common goal binding their actions into a single great pattern.

Without stockless production being incorporated in a rethinking of corporate strategy, the implementation of it becomes a copying of techniques. Copying is not good enough, and the implementation of techniques in a disjointed fashion is likely to be in conflict with established practices. Benefits will come from some of this, but the whole process is likely to be a matter of trying out fads.

STOCKLESS PRODUCTION IN OVERALL CORPORATE STRATEGY

The major thrust of a corporate strategy may be given in a very simple statement. For example, the Toyota Motor Company motto is, "Cars to love, the world over." One of the best known corporate mottoes in the United States is DuPont's: "Better things for better living through chemistry." These statements tell what the company will do, and they also tell what the company will refrain from doing.

Other companies have corporate goals which are not so succinctly stated, but which are nonetheless ingrained in the corporate culture over a period of many years. This feeling of shared purpose is embodied in many corporate cultures: Matsushita, Minnesota Mining and Manufacturing (3-M), General Motors, Hewlett-Packard, to name a few. One has the expectation of what these companies will be doing and something of the general direction in which they might head. A corporate strategy becomes the centerpiece of persuasion and acceptance in a company.

Having an operational strategy is very important in any company, large or small. An operational strategy points the direction in which physical and human assets will be headed. Cases of business failure studied in schools of business frequently emphasize the importance of operational strategies in different ways. For example, the collapse of the Curtis Publishing Company is one such case. What that story illustrates is a failure of entrepreneurial direction to adapt the operating strategy of the company to changing conditions. The original heirs and owners were unable to set a new course and hold with it.

Some companies at the very top appear to have a portfolio policy, which means that they will operate any business as long as the prospective returns are acceptable. This is seldom literally applied, but where it is, it is really a buy-and-sell policy for businesses, an attempt to sell trouble and buy opportunity, and that is inadequate for stockless production. From an operational point of view, corporate strategy has to deal with what must be physically done: markets, products, and how to provide them. Even if the organization goes bankrupt, the need for this does not always disappear. The trustees had to determine how to keep operating even after the failure of the Penn-Central Railroad.

In order to adopt stockless production in some form, a manufacturing company needs to have a stable overall operating strategy:

What markets to serve and how to serve them.

What are the products to be provided.

What must be produced.

How to financially enable all this.

Stockless production does not require the production function to be the central concern of corporate strategy. Every company is ultimately in the service business to the customer. However, the role of the production function needs to be clearly defined. The products cannot be made and delivered by magic. It may not be possible to clearly define the production strategy without a review of the overall corporate strategy.

Overall corporate strategy rarely changes radically. Sometimes a clear statement cannot be made because the internal politics of change do not allow clear statements which can polarize attitudes about the company at a time when change is needed. Some companies cannot handle revolutionary changes without self-destruction of the organization. Stockless production may be a part of a revolutionary change. It is necessary for top management to understand it and determine what overall strategy to pursue with respect to it.

For most companies, the adoption of stockless production has been motivated by fear—fear for survival, fear that competitors would surpass them in ability to deliver quality products at low cost, fear of what might happen if they did not adopt it. At such a time, the corporations as a whole undergo traumatic change in trying to assess the new reality. It is a hard job to set a futuristic course in such circumstances. Companies need to survive and prosper at a lower level of activity. They need to revise the nature of products offered. They need to approach customers in new ways.

For example, suppose the company considered it a strategic necessity to modify products to market them in the Far East, but that they must be produced in an American plant. That will stimulate a very serious review of how production is to be accomplished, focusing on the fundamentals of what must be done. (That is the reverse of the survival strategy that was seen by many Japanese companies, and it is a very illuminating exercise to consider it.)

Production strategy

The production strategy is only one piece of the overall corporate strategy. The company can build its strength by emphasizing research and development, by stressing access to markets and knowledge of customers, or in other ways. For some, such as cosmetics companies, the production process really just supports the overall feeling created by the marketing of the product. However, in any case it is useful to view the production process as an adjunct of overall strategy.

Exhibit 12–1 is one company's production strategy. Jidosha Kiki is primarily an automotive supplier of braking systems and suspension systems. They are affiliated with Bendix Corporation of the United States. The corporate strategies with respect to market, product, and research and development are not stated in the exhibit, but can easily be surmised.

Note at the top of the exhibit that the thought begins to develop with very simple statements of the overall objectives pertinent to production. As one moves down the page, the concepts of stockless pro-

EXHIBIT 12–1 Executive summary: Jidosha Kiki Company production strategy

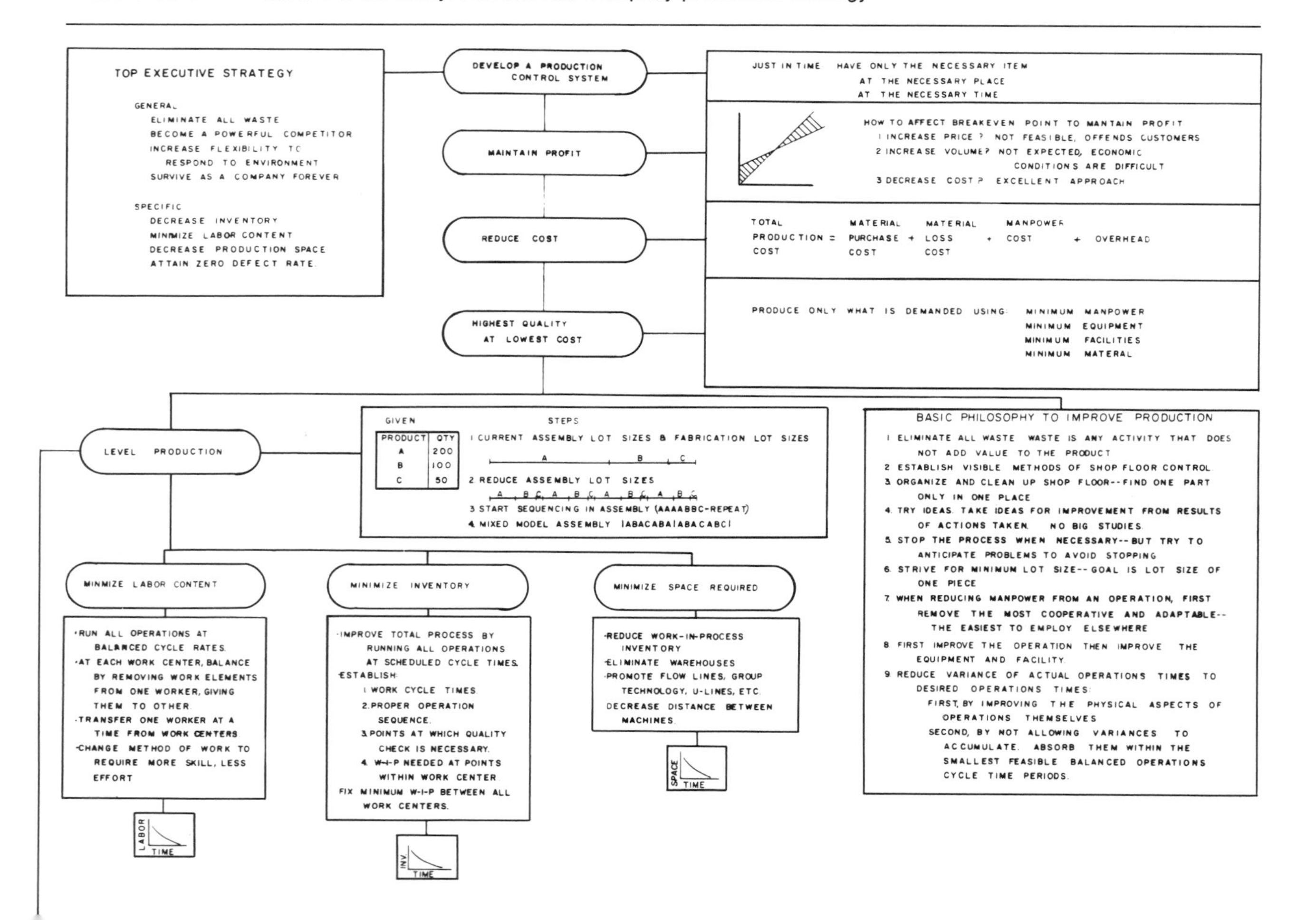

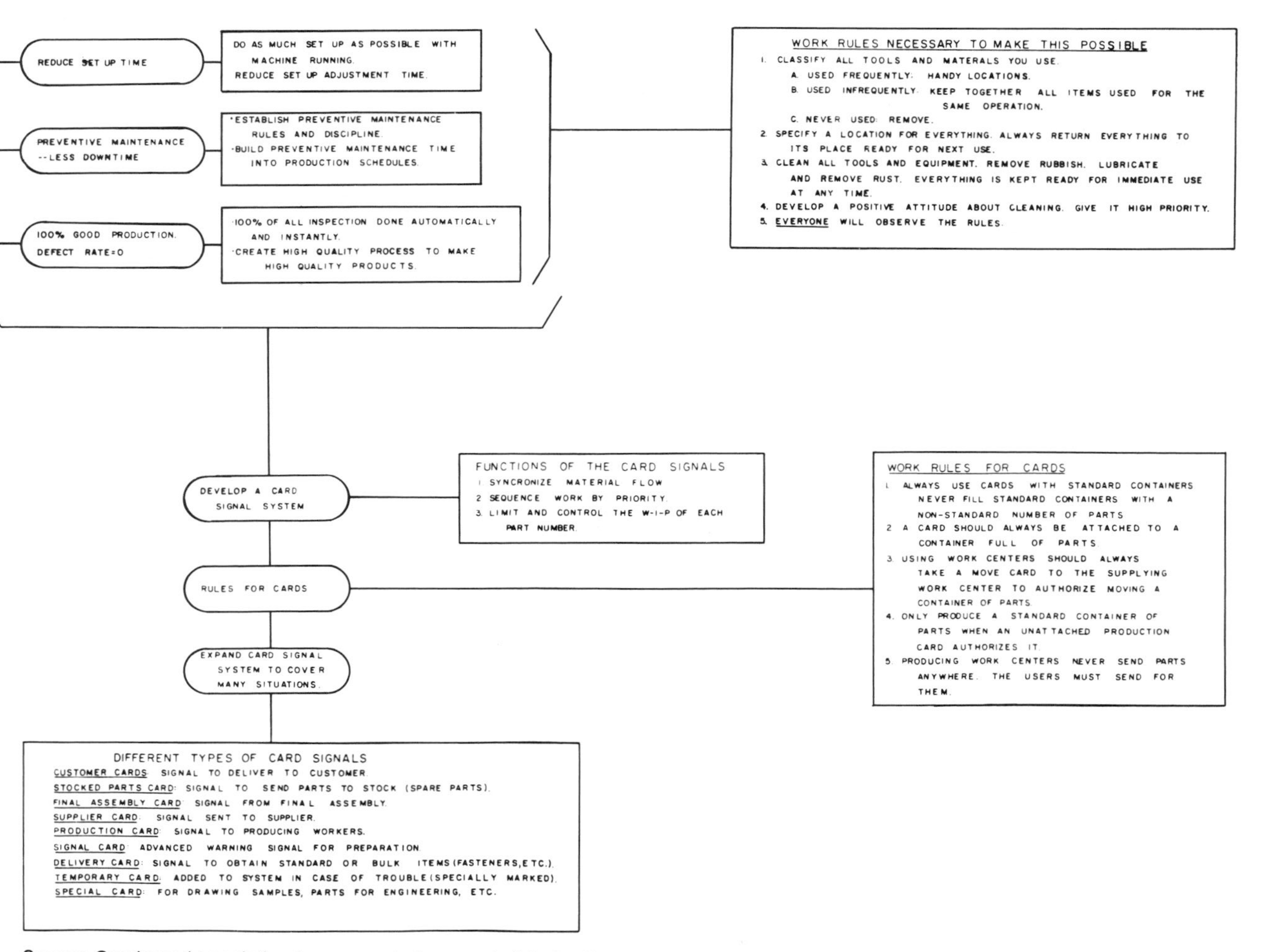

Source: Condensed translation from operator's manual, Jidosha Kiki Company, Ltd, with additions and deletions by J. Nakane and R. Hall.

duction become more and more specific, finally culminating in a description of the variations of card signals used in the Toyota Kanban System, which is the system Jidosha Kiki uses. (Other companies might not use it.) The sheet is an excellent summary of the methods of stockless production used in Japan and of the specific approach used by Jidosha Kiki. The executive objectives are translated into the approaches to work used in transforming the production floor.

Among the things not shown in Exhibit 12–1 is the directive from top management that the suppliers were not to be pressed to adopt the system themselves, and also that the adoption of the system was to be transparent to the companies who are their customers. The assumption of repetitive manufacturing is built in. Jidosha Kiki set out to greatly improve a standard repetitive manufacturing system by converting it to stockless production.

Note the goals in the box labeled "Top executive strategy." Survival and strength as an organization are the uppermost thoughts. Profit is a necessity to accomplish these goals, but it is not an overriding objective.

It is easy to relate stockless production to prospects of long-term profitability in any company, but profitability is not a very persuasive subject to the work force. The most compelling argument for both executives and the work force for all the known Japanese companies entering stockless production centered on the break-even point of the company. The financial objective was to decrease the break-even point. The informal agreement with all employees was that if you work to reduce the break-even point, and the total effort is successful, you are guaranteeing your future employment.

Developing the production strategy. Top management needs to determine long-term production goals and be prepared for a long-term program for stockless production. A strategy seems very simple once conceived and promulgated, but it is not a particularly easy thing to do when developing plant floor-level policies which relate to overall goals.

The production strategy starts from the basic corporate goals. The ideas of stockless production become embedded in it, and it comprises marketing, design, and corporate objectives into affirming the tasks the production function must do very well. Exhibit 12–2 shows one way to organize this thinking.

A production strategy which incorporates stockless production is a visionary thing. The goal is to automate, and it may not be possible to detail how the goal is to be reached. At the plant level, the approach is to begin working on problems in a directed way and persistently develop production toward the goal.

EXHIBIT 12–2 General framework of considerations for production strategy

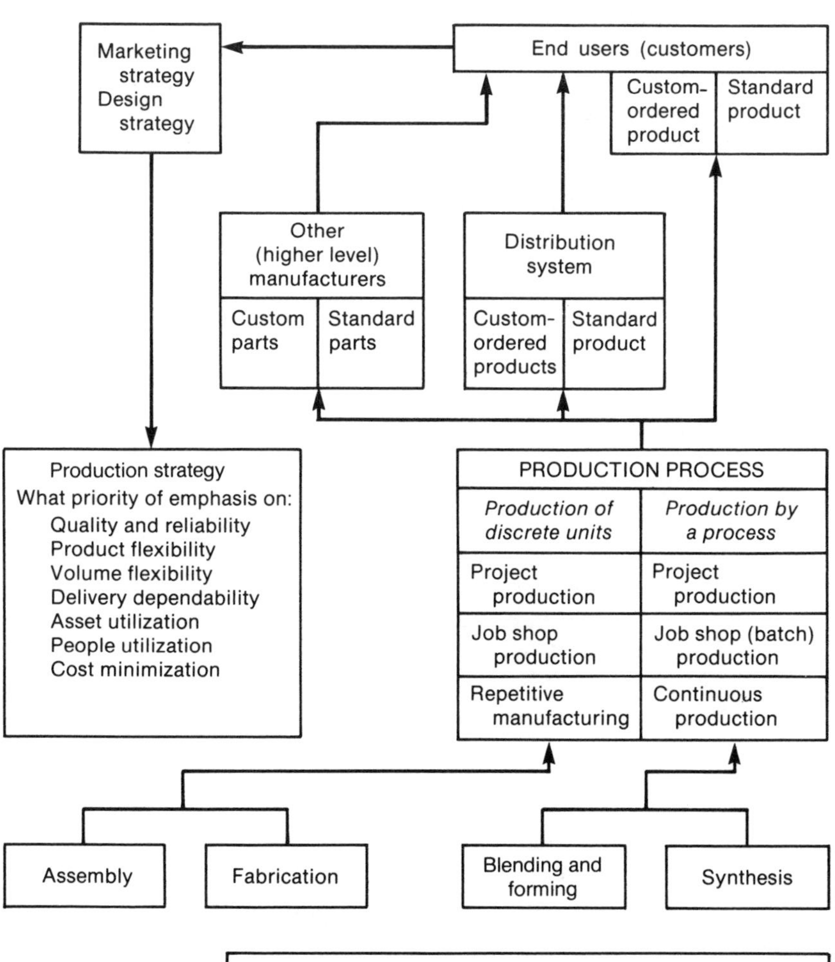

	Classification of parts and materials		
	Custom-designed production material families	Standard production material families	Other services and supplies
Type I Low cost suppliers	Tightly-linked supplier network	May or may not be tightly-linked	Other arrangements
Type II Special expertise suppliers	Often tightly-linked supplier network	Probably not tightly-linked	Other arrangements

The notion of production strategy does not have an extensive literature in the United States. Much of the thinking about it is summarized in a recent book by Wickham Skinner.[1] Production strategy means determining what must be done well in producing the product. Exhibit 12–2 lists several of the major elements to be considered in the strategy. Depending on the nature of the product and the market, emphasis may not be equal on all the goals listed. These considerations influence the thinking about production strategy and stockless production.

Quality and reliability. Improving quality in the sense of reducing defect levels should decrease cost, so it is not true that an emphasis on quality means sacrificing the emphasis on cost. However, the question of what constitutes defects is interesting in terms of what the market will accept. In some lines of business, the second-rate material can be sorted and sold to a second-rate market—ball bearings, for example. It is very tempting to think that improving quality does not pay as long as a product can be sold. However, suppose the same effort can result in a higher value for the product.

What is generally meant is something akin to closeness to tolerances required and the process capability needed to hold them. In stockless production, trading off defect levels for something else is a mode of thinking which should not be present. However, considering the range of tolerance capabilities required in production is a factor to consider. Producing parts which are at the limit of technical capability should not be done in the same plant in which a steady flow of parts is expected to come from processes which are easily capable of doing so.

Product flexibility. This can be interpreted two ways—products which have a large number of options and products which change very often. In any case, it comes down to whether the bills of material or the methods of production can be routinized. If the amount of custom engineering required is great, the production is a matter of operating a prototype shop, working to specifications different for many jobs.

The range of product necessary to manufacture is also a consideration. Is it necessary or desirable to select equipment which cannot be equipped for a flow of material? This question is a technical one, but often a very big problem. If the flow of parts cannot be untangled in a production operation as it stands, it is necessary to examine what is being done at the very foundation of practice. It is one thing to have a plant that can produce a large variety of different items if that is what the market wants; it is quite another to be unable to focus on the

[1] Wickham Skinner, *Manufacturing in the Corporate Strategy* (New York: John Wiley & Sons, 1978).

production of items produced so as to improve the methods by which they are made after these begin to be produced in volume.

One of the necessities is to separate the prototyping from other production if that prevents improvement. Another is to study how to reduce the time required to make a product change if frequent changes are necessary. For example, consider a plant making wiring harnesses for vehicles, appliances, or almost anything else. The wiring harness is one of the last items designed, and when there is an engineering change, it is one of the most frequent items affected. Therefore the production process should incorporate the means to make up new layout boards as quickly as possible.

This notion can be extended to other forms of production. If new molds must be introduced frequently, then a study of the mold-making process itself is in order. Perhaps the place where the ideas of stockless production could be very effective is in reducing the time required for this. It does not have to be regarded as a long lead-time. (Japanese companies are studying how to reduce the lead times in mold-and-die production so as to turn this into a competitive weapon.)

If a product is built with many options, the nature of the option process is itself a subject of study, not a given fact of life which necessarily prevents stockless production. As discussed earlier, the ideal way to handle options is to make the lead time for producing them so short that offering them to the market and incorporating them in production is not a reason to create a production process buried in inventory.

Volume flexibility. Every company which embarked on stockless production reported improved flexibility in this respect. This seems contradictory at first, because a period of fixed-rate production schedule to level material flows would seem to preclude sudden changes in output volume. However, the production rate for different products can vary greatly between the different periods of fixed schedule, as depicted in Exhibit 12–3. Therefore the production schedule can follow a seasonal trend as shown in that exhibit. The production schedule may be fixed for short periods of time; but it can be changed, as rapidly as material can be drawn through the pipeline, into a new fixed-schedule period, provided production work centers have been designed to respond to big changes each period.

As is always the case, the ability of production to follow changes in volume depends upon the lead times and capacity available. If a customer asks for a large increase in volume, the speed of response depends upon how fast material can be made available. There are only two ways to respond quickly: Have the material in inventory at some stage, or have the production capacity available to respond. The preferable way is to have capacity available if it can be done. There is less

LIBRARY ST. MARY'S COLLEGE

EXHIBIT 12–3 Seasonal production of a product with a fixed-period level schedule

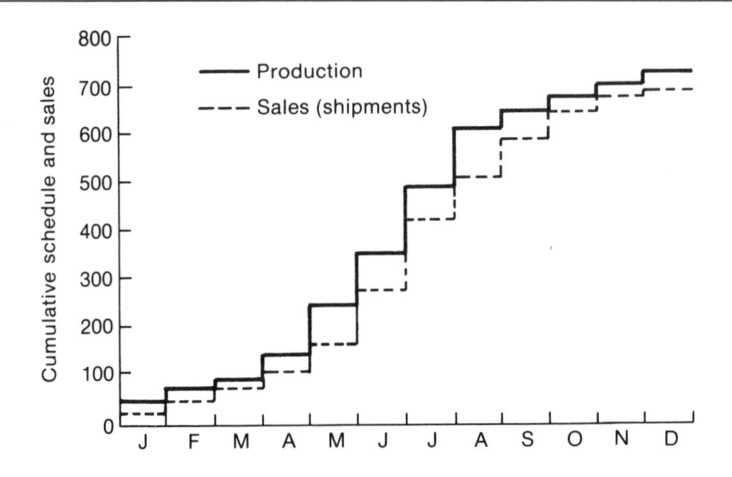

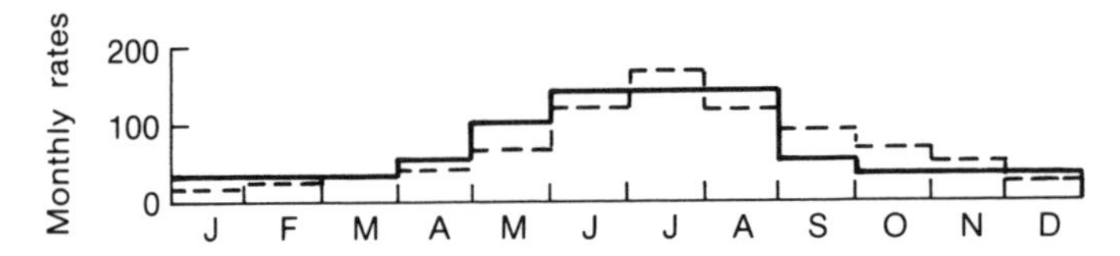

This is somewhat different from most charts. The interval represents each month (like a bar chart) rather than plotting points, as is more conventional. If the product is only one model in which some mix variations are permissible within the overall fixed-rate schedules, even more flexibility is possible.

risk of producing what is not really wanted. If it is necessary to build inventory to respond, that is a normal risk of doing business.

If capacity is not sufficient to produce at the rate needed at a seasonal peak, then material must be used to build products ahead of time and take some level of risk. Companies in stockless production report that their ability to respond is improved because actual production lead times are shorter and throughput times much shorter.

On the opposite side of the volume flexibility question is the problem of excess or obsolete material left on hand if sales suddenly turn down. The shorter the material pipeline, the less risk. In fact, if the entire production system contains only pipeline stock, this risk is confined to only what is in the pipeline.

The same logic applies in the case of style goods or items ordered by customers in only one or two major buys per year (Christmas ornaments, for example).

Typically, the department stores would order such items early in the season, based on a low-side estimate of impending sales, then reorder according to a late estimate of how much is needed to cover the season. Nothing in stockless production covers all the risks in stocking up for items of this type. However, if production can be made to respond more quickly and items are produced in shorter runs, the deliveries to customers might be possible in a lesser quantity. In short, the risks of matching production to sales might be lessened for both customer and producer.

Stockless production is in use for items which have a seasonal pattern in sales. Motorcycles are a case in point.

Delivery dependability. This is certainly a strong point of stockless production if the customers do not have volatile delivery demands. Those kinds of demands can only be met by having inventory somewhere if production capacity is inadequate to respond. Most companies using the fixed-schedule periods carry some finished goods inventory to service customers who have volatile demands, but they would carry as much inventory if they were not trying to speed flow through production. Concentrating on the amount of inventory carried just in case may overlook the benefits gained in many other ways by the improvements in the production process itself. Those benefits should be carefully considered before declaring that stockless production is impossible because the customers are unreasonable in delivery requests. One of the hardest aspects of stockless production philosophy is to learn to focus on the production process itself, and not on what is required as inventory just in case.

Asset utilization. The philosophy of this with stockless production differs from the conventional outlook, which usually is to concentrate on getting the most use from plant and equipment after it is in place. This is not just a simple matter of running it as much as possible. It begins with equipment selection. Use the simplest equipment to perform the needed task and modify it to fit into a pattern of almost continuous flow. That in itself tends to keep the cost of equipment down.

Likewise, the approach to material handling is to perform it in the simplest, most direct way, which means closing the distance between machines. The space released by this and by eliminating the need for stockrooms means less investment in brick and mortar. It also usually means less investment in material-handling equipment.

The notion of only running two shifts is sometimes troubling because it appears to decrease the amount of running time on equipment, and certainly there are operations which must be run around the clock—polymer extruders, for example. For much equipment, however, this is deceptive. By scheduling two shifts with preventive maintenance time in between, the uptime on equipment during the two shifts is improved. In addition, the problems of third shifts in plants are well known. People are less effective late at night; the problems of supervision increase, and, in general, the least-trained people are those who work that shift.

Several American plants have reported running two shifts with good results. Considering the improved quality of the process, the decreased downtime, and the absence of payroll costs for a third shift, the policy is generally a good one. The difference in production time on equipment can be as low as an hour or so. In addition, going back to the basic point of keeping work in process in tight control, there is only deception in having good, efficient equipment to make something that really is not needed.

Another reservation is whether extra tooling is needed in order to avoid lost production time if a tool should break. The great fear of production people is "What if a major die breaks?" This possibility always exists, and sometimes extra dies are kept for the purpose, but that is not the intent of stockless production. In fact, if people become accustomed to changing dies (or other attachments) frequently, there is less chance of breakage during a setup. The quality of the setup operations improves, and, if records are kept on tools and dies, the maintenance of them is decreased. The usual result of the stockless production program is an overall decrease in tooling expense and investment. The program stresses taking care of the equipment, which greatly assists in this. In fact, equipment which is carefully tended and modified may be better several years after it was purchased than when it was new.

An illustration of this comes to mind from a truck used for picking up truck tires by a tire-recapping service. George, the truck driver, drove the same truck for years. He changed the oil regularly, kept it tuned, and tended to the cause of every squeak. By the time George retired, the truck had over 400,000 miles, but had had no problem in service.

George's replacement considered it his job only to drive the truck, not to care for it, and his driving habits were rough. The truck was reduced to junk within six months—about 25,000 miles.

The same thing happens to production equipment of all types. The principle of asset preservation is a very powerful one in the problem of asset utilization, but not one often considered in corporate or production strategy, on the assumption that machine wear is an inevitable thing.

With a stockless production program, asset utilization may drop for some equipment, and that might be a problem, but, in general, the disciplines of stockless production all work in the opposite direction. In fact, attaining the effects of automation with the least equipment cost is the long-term objective, but it is easy to lose sight of the long-term effect with a quick calculation of theoretical utilization. This is an area to be very carefully evaluated before concluding that asset utilization would be too low. No company has reported such a result from stockless production.

In fact, Japanese auto companies, particularly Toyota, have much higher asset turnover ratios than their American counterparts. Part of this is explainable on the basis of depreciation policies, but all visitors to these plants remark on how evident is the ingenious use of simple equipment.

This comes home when considering the strategy of most people who start a manufacturing company on a financial shoestring. The ability to obtain used equipment and make it perform well is often the make-or-break factor. The same factor still applies after a company becomes large, but it is often forgotten. Strategy: Substitute skill and ingenuity for money.

People utilization. This is another strong point of stockless production. If programs to improve worker knowledge and versatility are accepted and put into action, the value of people rises dramatically. It is easy to fall into the habit of considering people, especially floor workers, as fixed quantities in terms of what they will do. In fact, if the program works, the job of every person is upgraded, and the skills they employ are increased.

Getting the staff focused on the operations also improves people utilization. Learn-as-you-do policies overcome the staff study sink-holes.

Cost minimization. Developing automation at minimum cost is the major thrust of stockless production, but it is not the only objective. A company may learn to handle mixes and changes at minimum cost as it becomes proficient, and this is important for products competing in markets where short life cycles are dominant. The ability to quickly get new products or new technology to market may be more important than making the learning curves as steep as possible. Emphasis in development of the system may shift more to engineering changes and start-ups.

Cost reductions from stockless production come from almost every category of cost, depending on where emphasis is placed as the program matures. In some of the companies, for example, a major thrust has been in energy saving. The dictum was "Be sure you know exactly what to make and how many to make before you turn a machine on." The lengthy setups and the scrap on the floor represent a waste of energy spent somewhere earlier in the total production process, and not just a waste of immediate time and labor.

Once a company has stockless production in grasp, a very good strategy debate can take place over how fast to move on into full automation. As mentioned earlier, one of the criticisms of Toyota is that they may overdo it on low-asset turnover and be later than necessary in introducing more advanced technology into production. The major thrust of this book has been that automation requires thought about much more than pure technological capability and cost, or it will be ineffective. However, to make stockless production most effective in the long run, it has to be a guide to the employment of advanced technology in production.

THE JOB SHOP, MRP, AND STOCKLESS PRODUCTION

Much of this book is based on the assumptions that repetitive manufacturing is possible and that manufacturers can convert to it. That is not possible for all of them. Anyone starting down the path toward stockless production first wants to study whether they are attempting something possible, although difficult.

It seems reasonable that most assembled products built in volume can be made by repetitive manufacturing. However, a large number of manufacturers do not produce items in volume. That alone does not mean that they cannot produce repetitively, but it lessens the likelihood. As we have seen, the mere existence of options does not define whether a company is really producing a large number of different items. One has to question whether the production of options requires the fabrication of a large variety of unique, irregular parts deep in the production process being considered.

Other companies have production operations that they consider processes. They formulate their own coatings, melt their own alloys, or swage their own tubing. Again, the mere fact that a production operation is a process does not mean that it could not be incorporated in stockless production.

Many companies have a combination of many kinds of production operations somewhere within them. Sometimes they are in separate plants, and sometimes they are under one roof. Analyze by beginning at final assembly and working backwards, asking what are the *valid* reasons why operations cannot be linked together by a pull system:

A final assembly schedule cannot be executed level enough to provide reasonable pull signals to the processes supplying it. Assembly requires fitting, trying, and testing and so it cannot be operated smoothly, and it is not foreseeable how this could be changed. The nature of the product demands it. It is not the result of poor quality or other conditions which can be improved.

Consumption of almost all parts is so low-volume and nonrepetitive that it is not reasonable to stock a pipeline which would feed final assembly and through which pull signals could be transmitted. (It might be possible to fabricate a few parts as demanded even if they are one of a kind, but for most this is too far out to be serious.)

It is necessary to commit a fabrication process to actual production in advance of any pull signal, based on actual demand which could be transmitted by any path from final assembly. This really means that it is impossible to reduce the setup times for the process by any means short of technology hoped for in the 21st century or beyond; and, as we have seen, imaginative use of current technology may find a way around that, so we have to beware of excuses.

The technology is such that quality cannot currently be improved enough, typically the case in integrated circuit yields or machining exotic materials.

Products must be produced as integrated lots all the way through production for reasons of quality records, certification, and so on. Pharmaceuticals are an example. However, some of these can be run almost continuously, so the exact nature of the restrictions must be considered.

By this means an analysis will show where stockless production is promising. Some of it is almost elementary, and many companies will see immediately that one division is an excellent candidate and another is not. Other times the analysis is not so elementary because the nature of production is tangled. The analysis can apply to a single company or to an industry. At some point, a stockless production company will want to analyze the industry of suppliers which feed it, or the industries of the companies they supply.

The idea is to analyze the company or industry on the basis of the production processes and material flows which exist (or could exist) within it, without obfuscation by current business relationships or management practices. It may be possible to change those.

A large company is likely to have several kinds of production, some of which are repetitive, or potentially that way, and some of which are job shop. That is certainly true of most industries making complex products. The strategy is to clearly separate the kinds of production so that each can be operated by the best possible methods for it.

A given plant may find that it could operate final assembly by a level schedule, and perhaps some of the operations feeding final assembly could be on a pull system, but several departments seem to be outside the scope of the program. Suppose that they operate as a job shop, using work orders on the floor. If this plant starts stockless production at final assembly, after a time it should resemble the bottom of Exhibit 12–4. Before stockless production, most companies operate by fabrication feeding parts by schedule into an inventory from which

EXHIBIT 12–4 Starting a job shop into stockless production

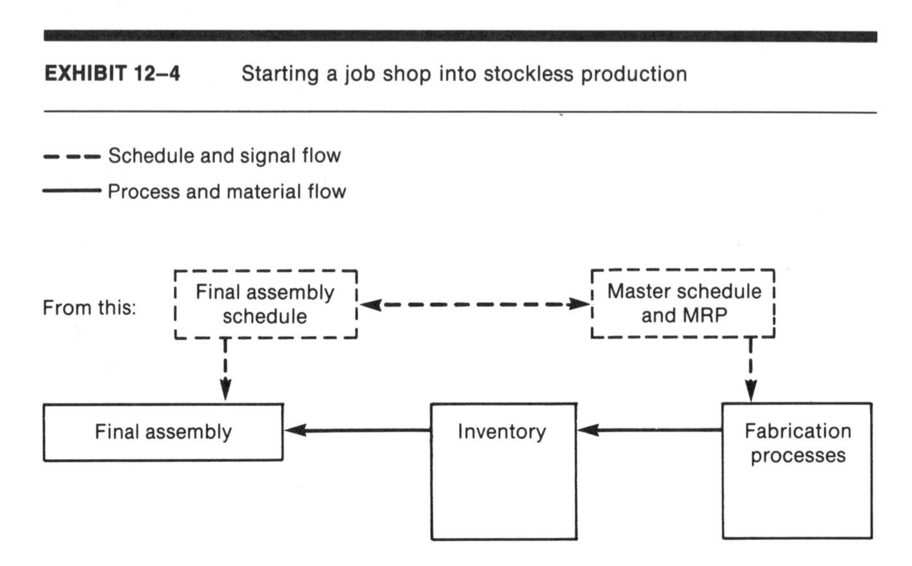

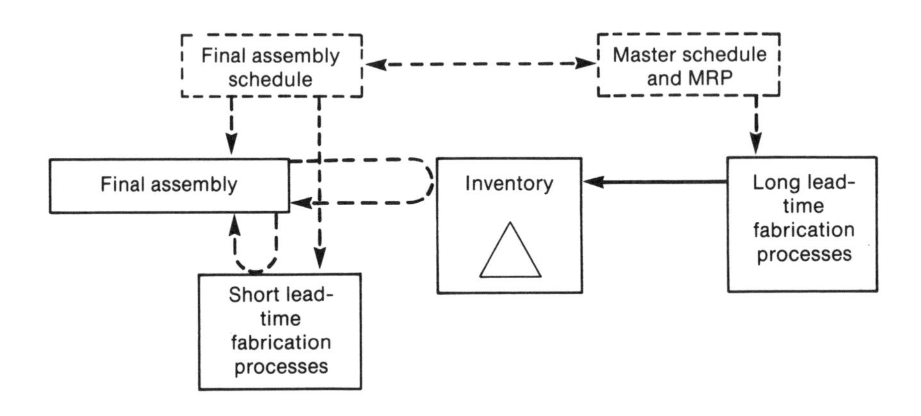

final assembly draws. After stockless production, final assembly may still draw some of the parts in the same way.

Most of the companies now on stockless production have pockets of fabrication which are still not on a clear pull system. Recall that some of them operated these small operations by a mini-MRP system done on a blackboard in the department. That is conceptually no different than pulling parts from the inventory of a supplier that operates on MRP.

Many general principles of stockless production apply to the job shop operations and plants, although some of the details of this book apply only to repetitive manufacturing, especially those concerning production planning and pull control. There is much more to improvement of job shops than good production planning and control, but that is an excellent start. Where inventory must be held in quantity and a push system used for scheduling and control, those things must be done well, and the literature of the American Production and Inventory Control Society has exhaustively covered the state of the art for this. However, there are many other practices approximating repetitive stockless production which can be applied to job shops:

Improve the quality of the processes by process capability studies and statistical quality control. Employ area control, failsafe methods, and immediate feedback as much as possible.

Reduce setup times. If done at low cost, it is hard to argue that this is useless for anyone.

Control the amount of work-in-process inventory. This is akin to exercising capacity control, and much has been written elsewhere about it. Disciplined counts and timing on the shop floor are the key to making any system work.

Revise and improve layout and housekeeping. Introduce smaller standard containers and standard material-handling methods as much as possible. The standard containers do much to improve count accuracy.

Fabricate materials in lot-for-lot quantities as much as possible. Do not consume capacity making anything which is unnecessary.

Attain visible control on the shop floor as much as possible. Organization of the shop floor helps any system to work properly. Markers, signs, and similar methods are useful. (Experiments in Japanese plants have shown that markers are useful in controlling the number of work orders at given points of the shop floor. For example, only 10 red markers might be available to put on work orders that have a bottleneck work center on the routing. If all 10 markers are out, another work order requiring one cannot be released until one of the markers out is retrieved along with the material which belongs to it.)

Do not allow material to sit unnecessarily in a stockroom. If possible, use the inventory system only as a control point to close one routing and start another, without physically storing the material.

Develop a preventive maintenance schedule. Allow time for it in the planned schedules for production.

Have foremen make up work plans, counterparts of what they would do for the schedule periods in a stockless production system.

Reduce what we have called variance in production processes as much as possible, and use the power of the systems available to help with this.

Variance refers to the causes of discrepancy between plan and actuality, and to the sources of uncertainty in the timing of work completed. We want to eliminate as much variance as possible, by physical and procedural changes, to reduce defects and unplanned downtime, then disperse the remaining variance so that it can be absorbed within the smallest units of time possible. This applies to the job shop as well as to the repetitive version of stockless production. Variance is a major root source of large queues on a job shop floor. Inability to balance load among work centers is another.

Absorption of variance in the repetitive version comes from leveling the schedule, production in short spurts at all operations, and balancing so that everything required (including residual variance) can be accomplished within a short cycle time in a confined location. The ability to run small lots and to split lots appears as if it can contribute greatly to this. One of the powers of optimized production technology (OPT) is that it stimulates plants to do this at points where it is desired to increase the flow rate of particular material.

Recall that in Exhibit 10–2, Sumitomo's wire plant operates as a job shop producing customer orders of different sizes. They made improvement through various programs to reduce variance so that they can actually run closely to what a scheduler has planned. Then they find it better to schedule each customer order as it arrives, backscheduling each order at each stop on its routing as an overlay on the orders already scheduled there. They also find it useful to split large orders so that long runs do not plug the flow of shorter orders. The wire plant is only a small shop and not as complex as many, but the principle is worth studying.

Material requirements planning is a method of backscheduling coming into widespread use. As generally employed, it backschedules by using offsetting lead times which allow for queues and variance on the shop floor. The only way to avoid overloads at the planning stage is by not overloading the master schedule and thus initiating the backscheduling. A great deal can be done to better integrate this planning tool with programs on the shop floor to obtain improved overall results.

However, the place to start is by examining how much of a plant that is now operated as a job shop can be operated by repetitive stockless production. It is simpler, and the potential for improvement is greater. Begin by shifting as much material acquisition as possible to the final assembly schedule rather than the MRP process used for fabrication. Actually, a number of companies do this for such things as packaging material, and occasionally for other short lead-time items. This seems natural for many Americans. The logic of it does not conflict very much with established job shop practice.

If the final assembly schedule can be leveled to where material can be issued in daily increments, the beginning of a course pull system can be established, and it may be possible to do much more than this by the methods already described. The final assembly process and those operations which can be tied directly to it can be developed as repetitive stockless production, and the remaining operations can be developed to excellence as a job shop.

The hazard for those who have long been used to operating by push-type material systems is that they will be caught up in the development of a mixed system which comes to be regarded as permanent, rather than give the attention necessary to develop the physical processes so that material flow can be smoothed further. Several of the first American experiments with a kanban system look very much like the system at the bottom of Exhibit 12–4, and the companies are fumbling to find ways to expand the system.

At some point, it is desirable to completely separate production processes that are hopeless to convert from a job shop method from those that can be converted to repetitive manufacturing. A plant cannot live by a mixed system, so in effect one plant becomes a supplier of another. The process of sorting this out requires an assessment of equipment selection, the potential for setup time reduction, and all the other elements of stockless production.

The point of all this is to carefully examine the details of the production processes themselves for improvement potential, and that has its own reward. It is better to separate the types of processes so that material flow is clear and improvement possibilities become clear than to enter into the complexity of trying to devise some type of mixed system on a permanent basis.

Every company has to experiment with its own production processes and think through how to reorganize them, and it is not easy. Several of the stockless production companies reported that one of the most difficult steps was to determine how to level the final assembly schedule and get started. It may not be possible to visualize how the production process will reorganize when first starting to work on it.

Even process industries can benefit from several of the principles of stockless production, so the review is not entirely in vain. For example, studying how to reduce the setup and adjustment of even a rolling mill can result in improved quality and a shorter scheduling sequence in working through a rolling schedule, although the details of a pull system of material control are foreign to such a process.

WORKING WITH SUPPLIERS: DEVELOPING AN INDUSTRY

Various companies contemplating stockless production may have three points of view, depending on their position within an industry:

1. *Assembly with a high percentage of parts fabricated themselves.* This company may feel that it has its destiny more in its own hands. The majority of processes for producing a product are under its direct control.
2. *Fabrication only.* This company is a supplier to other manufacturers, and may feel that leveling the schedule is extremely limited by the willingness of its customers to give it stable schedules.
3. *Assembly company with a low percentage of parts fabricated themselves.* Such a company may feel that it cannot much improve its own performance without beginning to work with suppliers early. Many of the production processes are not under their direct control.

The beginning point of developing an industry is visualizing it as a linkage of physical processes, not as a set of business relationships. Many suppliers may be separate supply plants owned by the assembly company. Some suppliers may be wholly dedicated to only one assembly process of one customer company, while, in other cases, the outside work centers must of necessity or of choice be shared with other assembly companies. In any case, the need to regard a feeding process as a supplier becomes more apparent when it becomes impossible to maintain close physical coordination between them, and different technical skills are required.

In this physical sense, each entity in an industry should be able to interact with a given set of markets, or customers, and a given set of suppliers by relatively simple and uniform methods. At the same time, the linkages between the outside work centers, which compose an industry, should be as tight as possible. The general method of developing such an industrial network is the same as developing stockless production in a given plant. Start solving one problem at a time to improve the production processes and streamline the material flow in order to see how to meld it together.

This approach to studying an industry is related to the argument whether an industry should be vertically integrated or not. Business and financial relationships certainly affect the way in which physical processes are likely to be linked, but the two subjects ought not be confused. A tight physical link for material flow may be established with a company whose ownership is totally independent. Likewise, the physical coordination may of necessity be more distant between two companies whose ownership is identical.

The risks of dealing with suppliers to form a closely linked physical supply net are easily understood. They would make any prudent management cringe, and yet, if no risk is taken, a great hope for major manufacturing improvement never is tried. The chances are that the success of suppliers participating in a program to improve quality and tightly link supply to customers will drive others to be receptive to the practice.

How to initiate and sustain communication to accomplish this with suppliers is a difficult question in the American context. While American companies are beginning to do this with selected suppliers, there is nothing in place comparable to the supplier associations of the major Japanese OEM companies. Certainly any American counterpart to those associations would be much more loose in nature, but it does not appear impossible to develop something. Major industries have long had committees for such things as setting standards. The Society of Automotive Engineers is an example. Similar organizations for developing the practices of stockless production seem possible if the need can be recognized and publicly accepted, and the legal problems considered. The Automotive Industry Action Group is an association which appears to be feeling its way into this role for the automotive industry in the United States. With development of stockless production, perhaps it will become accepted that being a supplier to the auto industry entails participation in that group if one is to know how to mesh into the network.

STRATEGY FOR IMPLEMENTING STOCKLESS PRODUCTION

There are limitless ways in which the implementation of stockless production can fail. Concentrating on trying to avoid all the mistakes that could cause failure or looking for the one path by which it is likely to be a success are not as likely to bring success as concentrating on doing those things which are likely to bring success. From what has so far been presented, it sounds as if the only way to implement is by starting with the corporate board of directors and proceeding with a careful strategy. In fact, many of the programs have not started at first with the top management. The lower-level management had to develop pilot projects as convincing demonstrations before top management could be persuaded. This happened, for example, at the Kawasaki Akashi Works in Japan.

The full participation of top management is required for the full effort necessary if stockless production is implemented as a complete program. However, in many cases, lower-level management must try to implement what they can without the support necessary for a complete revolution. A complete revolution will require a change in the corporate culture.

The Bendix Corporation has evaluated the ideas of stockless production to determine what kinds of activities they could expect to implement at the plant level without substantial revisions of patterns of behavior. Their list:

Just-in-time production methods (such as preventive maintenance).

Quality control at the source.

Uniform plant loads (in some plants).

Group-technology methods.

Setup-time reduction.

Pull systems of production control (where applicable).

How the production function is organized can strongly influence how a plant management can progress with stockless production. If plants are nearly independent with strong plant managers and the necessary staff responsibilities on site, the ability to do this is much enhanced. Unless the top management is strongly opposed, the plant management can do its own work without the cooperation of the corporate staff. Hewlett-Packard is an example of a company organized in this way. The plant and divisional managers can essentially play the role of top management in implementation.

On the other hand, if purchasing, engineering, industrial relations, or other functions are performed for a plant at a central corporate location, the plant management has a much more difficult time acting on its own. If they decide to try stockless production, they are limited to a small pilot project without corporate support.

As we have seen, stockless production tends to make people work closely together at the plant floor level. The more the organization is changed to put responsibility at the point of action, the more likely is success. The company which is lethargic in implementation is likely to be the one where so many people have to be involved in any basic action that every presentation of the project proposal ends with the recommendation, "It looks great! Why don't you contact ______?"

The most convincing argument for stockless production which can be presented to top management (or anyone else) is a demonstration. A demonstration overcomes reservations, doubts, and fears. It fuels the expectations, and it results in an accumulation of experience useful for extending stockless production to all applicable areas of the company. Stockless production is really only learned by doing it, so

EXHIBIT 12–5 Major strategic factors affecting conversion to stockless production

Factor	*Possible action*	*Comments*
1. Sales volume	Increase/decrease/hold	Tempting to disregard it if sales are increasing. —Usual stimulant—stagnant sales.
2. Production versus capacity	Near capacity/below capacity	Need extra capacity at start of conversion. —Hard for production to concentrate on it if all energy goes into getting out product. Build new plants to start up on stockless production? —Possible, but what about converting rest of organization?
3. Competition based on cost	Expect increase/decrease	Fear of low-cost competition a powerful motivator. —But is success of company really based on other factors, such as strong research and development, strong advertising, and so forth? —Can company afford to drop low-cost products, as Hewlett-Packard did with calculators? —Cost competition can come to the higher-cost end of a product line, too.
4. Profit and cash flow	Positive/negative Need increase or hold Afford temporary decrease?	How much time do we have? Must we go for large cost reductions quickly? —If so, it may be that we cut the very people needed to make stockless production work. —Are we willing to take a lean year or two to get this going?
5. Quality of product	Must increase/hold/decrease Number of defects Seriousness of defects	If an immediate decrease of defects to the field is needed, that sets the initial moves with stockless production.
6. Quality of process (process control)	Tighten/relax/hold variance in processing	Requires thinking through the current status of process control versus what is needed. Do we need to give very great emphasis to this in the beginning? —Do we know the variances (tolerances) of production processes now? —Are defect levels so high that they must be reduced in many operations before we can seriously reduce inventory? Is the plant full of rework loops?

EXHIBIT 12–5 *(continued)*

Factor	*Possible action*	*Comments*
		—Do we need to reevaluate or revise a lot of specifications to understand what kind of process control is really required to allow immediate feedback (or immediate inspection) systems to work for many quality characteristics more technical than just completeness of operations? —Do we need to start from scratch with a preventive maintenance program?
7. Complexity of design	Must it increase? Can we decrease or hold?	Stockless production demands evaluation of producibility of design. —Can we quickly dispense with many obsolete parts to simplify keeping one part in one place? —Is it necessary to revise designs to reduce the proliferation of parts? —Can we refine designs to make production operations simpler? (Do we have a strong value-analysis program oriented to production, not just to getting lower bids on purchased parts?)
8. Rate of introduction of new products/models	Increase/decrease/hold	This has a great deal to do with the necessary flexibility required in production. —Do we need to redevelop the organization of design engineering to smooth introductions into production? —What is the nature of changes? Mostly cosmetic? Involve great ranges in production requirements? —Is it possible to greatly refine the process of starting up new products or models? Or do many of them require new equipment, not just new tooling?
9. Number of engineering changes	Reduce or hold? Increase likely?	Do we need to reexamine this thoroughly? —Is it possible to avoid many changes by better start-ups? Are many of them trivial? —Can Production Control keep the bill of material for products in production? —Can most of them be grouped?

	Factor	Assessment	Comments
10.	Factory focus	Increase/decrease/hold	This is a difficult factor. It is much easier to see how to proceed with stockless production if products of a plant are not a tangled mix of everything. —Reassign products among plants? —Break plant into clear divisions, plants within plants. —Assign some parts or products to be made by suppliers? All this is an extension of the process of improving housekeeping and layout to clarify problems. It may not be possible to clearly see what major decisions on factory focus must be made until the housekeeping preparation is underway.
11.	Type of production (part of factory focus)	Acceptable, or need to restrict	Important not to accept the current situation without reflection. —Can job shop operations be converted to repetitive? Which ones cannot? —Do we have a high percentage of fabrication? It is tempting to think that stockless production will be easier in a plant if it is mostly an assembly plant. However, that requires work to evaluate or improve the efficiency of suppliers and the transport of material from them. That may not be easy. (Stockless production covers the total *industry* process.)
12.	Current degree of automation		This may or may not be a big advantage. If the current mode of automation promotes a level schedule and/or flexible production, it may help. We usually have a better idea of process capabilities when operations have been recently studied or automated. However, it may be that automated equipment needs to be removed and a new start made toward automation. —Is equipment intrinsically difficult to set up? —Is it difficult to maintain? —Do we have a strong internal organization to revise or develop equipment?
13.	Degree of system change	The more severe the change, obviously the more difficulty	This gets into the difference between an evolution and a revolution. —Do we need to change from a job shop system? —Is a major logistics change needed? —Will there be major plant relocations? —Must we make major changes in personnel policies? —Do we need to drastically change organizational methods of operating? —Are we a multiplant company? —Must there be great revision of operations conducted at a central headquarters and at plant level?

EXHIBIT 12–5 *(concluded)*

Factor	*Possible action*	*Comments*
14. Relations with suppliers		The question is how major the changes must be. —Quality from suppliers must improve. —Need to establish methods for determining the causes of poor supplier quality, ways to reduce cost, and ways to simplify operations. —Is the company really dependent for technology on many of its suppliers? If so, that establishes the nature of a relationship that may not be easy to change. —How difficult will it be to reduce the number of suppliers if that is determined to be desirable? —Is a search for new suppliers needed? —Must many suppliers be developed in skill to perform to the standards of stockless production?
15. People problems	Many changes/few changes	The quality and morale of people are the key to making anything work. —Do we need to set a new policy of guaranteed employment or guaranteed future for employees? —Do we currently have excess employees? How do we establish the core group we can afford to retain? Simplification of production and management will lead to decreased need for both hourly and salaried personnel. Example: What happens to inventory control personnel if practically nothing is left in the storerooms? —Do we have experienced people? People capable of floor leadership as implementation teams? —Are foremen capable of leading people through the changes? —Do we need major revisions in performance measurement and pay practices? How will this be accepted?

Note: This table is incomplete because there is no end to the kinds of open questions which may be evaluated in the course of determining strategy. It is not possible to set the perfect strategy, but rather, the objective of a strategy review is to avoid the kinds of major oversights which can lead to wasting a lot of time—or experiencing disastrous failure. The strategy questions never end; the results of progress in improving plant floor activity constantly suggest more. The more that is done, the more one can see to do.

the proper form of organization for it is one which encourages direct action and learning from the mistakes.

How extensive a change in corporate culture is necessary to implement full-scope stockless production at most corporations? What is corporate culture, and how is this done? Corporate culture is a pattern of thinking and a set of shared values. It can be changed, but it is hard to change it by merely invoking the change. It is changed by changing the things which people do, slowly changing the daily patterns of interaction by which people do their jobs. It has been done. All the companies who have converted to stockless production have said that it was not a modest extension of the existing corporate culture, but a major change in a way of life. They did it by concentrating on developing a propensity for continuous improvement—by going to the plant floor and starting. The culture change needed in the beginning is to stimulate people to do this.

At the conclusion of a study trip to Japan in the summer of 1982, several American manufacturing executives were asked to write what they needed to do to implement stockless production in their companies. The most apparent conclusion was that those who chose to write about this from the perspective of changing the physical activity on a plant floor had a far easier time than those who took the perspective of a corporation with centralized staff functions and interdependent plants. It is far harder to understand how to change the practices of a corporate accounting, engineering, purchasing, or scheduling group. The way to make the change is to bring responsibility as close as possible to where the physical action is. Then it becomes more clear, and the corporation changes culture by people seeing what to do and doing it. People of all cultures play baseball almost the same way, but not necessarily with the same gusto or theatrics.

However, reorganizing at the corporate level for stockless production, is not a simple matter of getting everyone closer to the plants. For example, consider what an engineering department must do well in order to make design engineering an imaginative, integrating activity:

Know the needs of the customers and the advantages and disadvantages of current products in field service.

Know the relevant technology so as to bring the possible to reality.

Know the capability of the plants.

Know what parts and materials suppliers can deliver, and the quality, reliability, and performance in service.

Participate in the production start-up of new designs.

Minimize routine engineering changes.

This is a very integrative set of activities, not a highly specialized one. It points in the direction of having design teams in which communication in many directions is a fundamental need. An engineering department which is buried in bureaucratic forms of communication cannot perform these functions very well. Many engineering departments may need to reconstitute themselves to perform in this way. In the process, corporate management may need to deal with the problem of extra people who have been added because everyone has been working ineffectively.

The engineering department is only one of several corporate-level activities which may need redirection in order to perform adequately in a world in which lead times have been drastically reduced. However, the needs of stockless production are only one set of needs to which design engineers must pay attention.

It is harder to determine how to reorient corporate-level activities to support it, but the principles of stockless production implementation there are the same as at the plant level. Revise the activities to support what has to be done at the point of action in the plants and with suppliers.

The number of issues to consider in developing a strategy of implementation can be lengthy, as illustrated in Exhibit 12-5. However, the best advice is to work at keeping it simple.

Professor Jinichiro Nakane has had considerable contact with the major Japanese practitioners of stockless production. When asked to summarize the most important points of what they do in few words, he responded:

The simplest way is best. Study the physical problems carefully and try to solve them by simple means before turning to more complex ones. Simple solutions to problems are more likely to succeed.

The longest way around may be the shortest way home. Take a step at a time and be patient. People can only respond to a limited number of changes at one time. Make big improvements through a series of small improvements.

Index

T

This book has been set Linotron 202, in 10 and 9 point Caledonia, leaded 2 points. Chapter numbers are 24 point Caledonia and chapter titles are 18 point Caledonia Bold. The size of the type page is 27 by 47 picas.